Birkhäuser

Trends in Mathematics

Research Perspectives Ghent Analysis and PDE Center

Volume 7

Series Editor

Michael Ruzhansky, Department of Mathematics, Ghent University, Gent, Belgium

Research Perspectives Ghent Analysis and PDE Center is a book series devoted to the publication of extended abstracts of seminars, conferences, workshops, and other scientific events related to the Ghent Analysis and PDE Center. The extended abstracts are published in the subseries Research Perspectives Ghent Analysis and PDE Center within the book series Trends in Mathematics. All contributions undergo a peer-review process to meet the highest standard of scientific literature.

Volumes in the subseries will include a collection of revised written versions of the communications or short research announcements or summaries, grouped by events or by topics. Contributing authors to the extended abstracts volumes remain free to use their own material as in these publications for other purposes (for example a revised and enlarged paper) without prior consent from the publisher, provided it is not identical in form and content with the original publication and provided the original source is appropriately credited.

Roland Duduchava • Eugene Shargorodsky •
George Tephnadze
Editors

Tbilisi Analysis and PDE Seminar

Extended Abstracts of the 2020–2023 Seminar Talks

Birkhäuser

Editors
Roland Duduchava
Institute of Mathematics
University of Georgia
Tbilisi, Georgia

Eugene Shargorodsky
Department of Mathematics
King's College London
London, UK

George Tephnadze
Institute of Mathematics
University of Georgia
Tbilisi, Georgia

ISSN 2297-0215 ISSN 2297-024X (electronic)
Trends in Mathematics
ISSN 2948-1724 ISSN 2948-1732 (electronic)
Research Perspectives Ghent Analysis and PDE Center
ISBN 978-3-031-62893-1 ISBN 978-3-031-62894-8 (eBook)
https://doi.org/10.1007/978-3-031-62894-8

This book is published under the imprint Birkhäuser, www.birkhauser-science.com by the registered company Springer Nature Switzerland AG
The registered company address is: Gewerbestrasse 11, 6330 Cham, Switzerland

Preface

The aim of this volume is to present new developments and ideas in various fields of *Analysis and Partial Differential Equations*, including harmonic analysis, operator theory, function spaces, inequalities, integral equations, and applications. It provides a comprehensive review of some latest results and ideas in these rapidly expanding fields. It also contains some open problems, which we hope will inspire further research. The book contains 21 research articles and covers a broad spectrum of topics and applications.

The contributing authors have given talks at, have participated in, or are related in some way to *Tbilisi Analysis & PDE seminar* and *Tbilisi Analysis & PDE workshop*. The seminars are held bi-weekly since 2020, and the workshop will be organised annually following the first workshop that was held at the University of Georgia, Tbilisi, on 30 August 30–2 September 2023. Participants of the workshop and the seminars have enjoyed inspiring talks by *Maria Esteban*, *Pavel Exner*, *Hans Feichtinger*, *Gerd Grubb*, *Ari Laptev*, *Volker Mehrmann*, *Lars-Erik Persson*, *Michael Ruzhansky*, *Kristian Seip*, *Mikhail Sodin*, and by many others.

The Seminars and Workshops advance mathematics research in Georgia, enhance its visibility, and foster collaboration between mathematicians from the region and from other countries. They aim to offer a general perspective of the current mathematics research in Georgia. Georgia has a strong tradition in Analysis, Integral and Partial Differential Equations (PDEs), and the seminar organisers continue efforts to inspire the young generation in Georgia to follow the traditions of the Georgian Mathematical School. We are particularly happy that a good proportion of the papers in this volume present research by young mathematicians in Georgia.

This series is supported by our partner—the Ghent Analysis and PDE Center, which has provided us with a framework to publish this book in the series: *Research Perspectives Ghent Analysis and PDE Center*. We thank the Editor in Chief of the series, Prof. Dr. Michael Ruzhansky, for his support, without which of our project would not have been possible.

We would like to extend our cordial thanks to our colleagues *Lars-Erik Persson* (UiT - The Arctic University of Norway), *Michael Ruzhansky* (Ghent Analysis and PDE Center), and *Duvan Cardona* (Ghent Analysis and PDE Center), who have served as reviewers for this volume, for all the unconditional support they have given to this series. Also, we thank all our colleagues involved in the organisation of our past seminars and workshop, namely, *Medea Tsaava* (The University of Georgia), *Davit Baramidze* (The University of Georgia), *Giorgi Tutberidze* (The University of Georgia), and *Zurab Vashakidze* (The University of Georgia). We also thank *Kunda Kambaso* and *Daniel Jagadisan* from Birkhäuser-Springer for their help.

Tbilisi, Georgia — Roland Duduchava
London, United Kingdom — Eugene Shargorodsky
Tbilisi, Georgia — George Tephnadze
April 2024

Contents

Continuous Inequalities: Introduction, Examples and Related Topics 1
Joachim Jørgensen Ågotnes, Ludmila Nikolova, Lars-Erik Persson, and Sanja Varošanec

Approximation by Vilenkin-Nörlund Means in Lebesgue Spaces 11
Nino Anakidze, Nika Areshidze, and Lasha Baramidze

On Divergence of Fejér Means with Respect to Walsh System on Sets of Measure Zero 21
Nika Areshidze, Lars-Erik Persson, and George Tephnadze

Martingale Hardy Spaces and Some Maximal Operators Associated with Walsh-Fejér Means 31
Davit Baramidze

Generic Bessel Potential Spaces on Lie Groups 43
Duván Cardona, Roland Duduchava, Arne Hendrickx, and Michael Ruzhansky

The Finite Hilbert Transform Acting in Rearrangement Invariant Spaces on $(-1, 1)$ 55
Guillermo P. Curbera, Susumu Okada, and Werner J. Ricker

The Banach Gelfand Triple and Its Role in Classical Fourier Analysis and Operator Theory 67
Hans G. Feichtinger

A Note on a Frictional Unilateral Contact Problem in Nonlinear Elasticity 77
Joachim Gwinner

Maximal Noncompactness of Singular Integral Operators on L^p **Spaces with Power Weights** 87
Oleksiy Karlovych and Alina Shalukhina

A Remark on Piecewise Linear Interpolation of Continuous Fourier Multipliers 99
Oleksiy Karlovych and Eugene Shargorodsky

Banach Algebras of Convolution Type Operators with PQC Data 109
Yuri Karlovich

Integrability and Convergence of Trigonometric Series and Fourier Transforms 121
Elijah Liflyand

Commutators of Calderón–Zygmund Operators in Grand Variable Exponent Morrey Spaces, and Applications to PDEs 131
Dali Makharadze, Alexander Meskhi, and Maria Alessandra Ragusa

Symmetric Stein–Tomas, and Why Do We Care? 143
Rainer Mandel and Diogo Oliveira e Silva

On Solonnikov Parabolicity of the Evolution Anisotropic Stokes and Oseen PDE Systems 153
Sergey E. Mikhailov

On Convergence and Divergence of Fourier Series and Féjer Means with Applications to Lebesgue and Vilenkin-Lebesgue Points 161
Nato Nadirashvili and Tural Nasibov

On Generalized Sharpness of Some Hardy-Type Inequalities 171
Lars-Erik Persson, Natasha Samko, and George Tephnadze

Interaction Problems for n-Dimensional Dirac Operators with Singular Potentials 183
Vladimir Rabinovich

Convergence and Summability in Classical and Martingale Hardy Spaces 191
George Tephnadze

Modulus of Continuity and Convergence of Fejér Means of Vilenkin-Fourier Series in the Variable Martingale Hardy Space $H_{p(\cdot)}$ 201
George Tephnadze and Ferenc Weisz

On Unconditional Convergence of Fourier Series with Respect to General Orthonormal Systems 211
V. Tsagareishvili, G. Tutberidze, and G. Cagareishvili

Continuous Inequalities: Introduction, Examples and Related Topics

Joachim Jørgensen Ågotnes, Ludmila Nikolova, Lars-Erik Persson, and Sanja Varošanec

Abstract Classical inequalities in a measure space are usually described by involving finitely many functions $f_1, \ldots, f_N$, $N = 2, 3, \ldots$. In this paper we give an introduction and several examples when such inequalities can be given with infinitely many functions f_s involved, where index s can even be taken from another measure space. Such a development was also inspired when developing an interpolation between families (continuously many) of Banach spaces.

Keywords Inequalities · Continuous inequalities · Hölder's inequality · Minkowski's inequality · Bellman's inequality · Popoviciu's inequality · Refinement · Interpolation between families of Banach spaces

2010 Mathematics Subject Classification 26D15, 26D10, 26D20, 39B62, 46E27

J. Jørgensen Ågotnes
Department of Computer Science and Computational Engineering, UiT - The Arctic University of Norway, Narvik, Norway
e-mail: joachim.j.agotnes@uit.no

L. Nikolova
Department of Mathematics and Informatics, Sofia University, Sofia, Bulgaria
e-mail: ludmilan@fmi.uni-sofia.bg

L.-E. Persson (✉)
Department of Computer Science and Computational Engineering, UiT - The Arctic University of Norway, Narvik, Norway

Department of Mathematics and Computer Science, Karlstad University, Karlstad, Sweden

S. Varošanec
Department of Mathematics, Faculty of Science, University of Zagreb, Zagreb, Croatia
e-mail: varosans@math.hr

R. Duduchava et al. (eds.), *Tbilisi Analysis and PDE Seminar*,
Research Perspectives Ghent Analysis and PDE Center 7,
https://doi.org/10.1007/978-3-031-62894-8_1

1 Continuous/Family Hölder-Type Inequalities

The integral Hölder inequality reads: *if* p *and* q *are positive numbers such that* $\frac{1}{p}+\frac{1}{q}=1$ *and if* f *and* g *are non-negative functions on the measure space* (Y, ν), *such that* $f \in L_p(Y)$, $g \in L_q(Y)$, *then*

$$\int_Y f(t)g(t)\,d\nu(t) \leq \left(\int_Y f^p(t)\,d\nu(t)\right)^{1/p}\left(\int_Y g^q(t)\,d\nu(t)\right)^{1/q}. \tag{1.1}$$

The Hölder inequality (1.1) can obviously be formulated also for $n = 3, 4, 5, \ldots$ functions involved. However, to get a hint how to generalize it with infinitely many functions involved, we rewrite (1.1) as an inequality between two geometric means:

$$\int_Y f^{1/p}(t)g^{1/q}(t)\,d\nu(t) \leq \left(\int_Y f(t)\,d\nu(t)\right)^{1/p}\left(\int_Y g(t)\,d\nu(t)\right)^{1/q}.$$

The continuous version of the Hölder inequality (as an inequality between generalized geometric means) reads:

Theorem 1.1 *Let* u *and* v *be weight functions on the measure spaces* (X, μ) *and* (Y, ν), *respectively, such that* $\int_X u(x)\,d\mu(x) = 1$. *Let* f *be a positive function on* $X \times Y$ *and measurable with respect to the measure* $\mu \times \nu$. *Then*

$$\int_Y \exp\left(\int_X \log f(x,y)u(x)\,d\mu(x)\right)v(y)\,d\nu(y)$$
$$\leq \exp\left(\int_X \log\left(\int_Y f(x,y)v(y)\,d\nu(y)\right)u(x)\,d\mu(x)\right). \tag{1.2}$$

Corollary 1.2 *Let* $w_1, \ldots w_m \geq 0$ *be real numbers and let* $u \geq 0$, $p > 0$, $a_i > 0$ $(i = 1, 2, \ldots, m)$ *be functions on* X *such that* $\displaystyle\int_X \frac{u(x)}{p(x)}d\mu(x) = 1$ *and* a_i^p *are measurable on* X. *Then*

$$\sum_{i=1}^{m} w_i \exp\left(\int_X \log a_i(x)u(x)\,d\mu(x)\right)$$
$$\leq \exp\left(\int_X \log\left(\sum_{i=1}^{m} w_i a_i^{p(x)}(x)\right)\frac{u(x)}{p(x)}d\mu(x)\right).$$

Remark 1.3 Putting in Corollary 1.2: $m = 2$, $w_1 = w_2 = 1$, $p(x) = 1$ we get the following result, which can be described as superadditivity of geometric means:

$$\exp\left(\int_X \log a_1(x)u(x)\,d\mu(x)\right) + \exp\left(\int_X \log a_2(x)u(x)\,d\mu(x)\right)$$
$$\le \exp\left(\int_X \log\left(a_1(x) + a_2(x)\right)u(x)\,d\mu(x)\right).$$

Example 1.4 Let $N = 2, 3, \ldots$, $X = [0, 1]$, $d\mu(x) = dx$, $u \equiv 1$, $p_i > 1$ for $i = 1, 2, \ldots, N$, with $\sum_{i=1}^{N} \frac{1}{p_i} = 1$, $f(x, y) = f_1^{p_1}(y)$, $0 \le x \le \frac{1}{p_1}$, $f(x, y) = f_2^{p_2}(y)$, $\frac{1}{p_1} < x \le \frac{1}{p_1} + \frac{1}{p_2}, \ldots$, $f(x, y) = f_N^{p_N}$, $\sum_{i=1}^{N-1} \frac{1}{p_i} < x \le 1$. Then (1.2) reads:

$$\int_Y \prod_{i=1}^{N} f_i(y)v(y)\,dv(y) \le \prod_{i=1}^{N}\left(\int_Y f_i^{p_i}(y)v(y)\,dv(y)\right)^{1/p_i},$$

i.e. it is a standard form of the Hölder inequality involving N functions.

There exists some refinements of several of the classical inequalities. Such refinements can also be proved for several continuous inequalities. Our refinement of the continuous Hölder inequality in Theorem 1.1 reads:

Theorem 1.5 *Let $f(x, y)$ be a positive and measurable function on $(X \times Y, \mu \times \nu)$ and let $u(x)$ and $v(y)$ be weight functions on X and Y, respectively, such that $\int_X u(x)\,d\mu(x) = 1$ and $\int_Y f(x, y)v(y)dv(y) > 0$ μ-a.e. Moreover, let (Z, dz) be a measure space and $\alpha(z, y)$ be a non-negative integrable function on $Z \times Y$ such that $\int_Y \alpha(z, y)f(x, y)v(y)dv(y) > 0$ μ-a.e and $\int_Z \alpha(z, y)\,dz = 1$, for $y \in Y$. Then the following refinement of continuous form (1.2) of the Hölder inequality holds:*

$$\int_Y \exp\left(\int_X \log f(x, y)u(x)\,d\mu(x)\right) v(y)\,dv(y)$$
$$\le \int_Z \left[\exp \int_X \log\left(\int_Y \alpha(z, y)f(x, y)v(y)\,dv(y)\right) u(x)\,d\mu(x)\right] dz$$
$$\le \exp\left[\int_X \log\left(\int_Y f(x, y)v(y)dv(y)\right) u(x)\,d\mu(x)\right].$$

2 Continuous/Family Minkowski-Type Inequalities

The main result in this case reads:

Theorem 2.1 *Let $f(x, y)$ be non-negative and measurable on $(X \times Y, \mu \times \nu)$ and let $u(x)$ and $v(y)$ be weight functions.*

(a) If $p \geq 1$, then

$$\left(\int_Y \left(\int_X f(x, y)u(x)\,d\mu(x)\right)^p v(y)\,d\nu(y)\right)^{\frac{1}{p}} \leq \int_X \left(\int_Y f^p(x, y)v(y)\,d\nu(y)\right)^{\frac{1}{p}} u(x)\,d\mu(x). \tag{2.1}$$

(b) If $0 < p < 1$ and

(i) $\int_X \left(\int_Y f(x, y)v(y)\,d\nu(y)\right)^p u(x)\,d\mu(x) > 0 \quad \mu\text{-a.e. and}$
$\int_Y f(x, y)v(y)\,d\nu(y) > 0 \quad \nu\text{-a.e.},$
then the reverse inequality in (2.1) holds.

If $p < 0$, the above-mentioned assumptions (i) and the additional one

(ii) $\int_X f^p(x, y)u(x)\,d\mu(x) > 0 \quad \nu\text{-a.e.}$
hold, then the reverse inequality in (2.1) holds.

Example 2.2 Let $X_1, \ldots, X_N$ be a partition of X, $\int_{X_i} u(x)\,d\mu(x) = \alpha_i$, and $f(x, y) = \frac{f_i(y)}{\alpha_i}$, for $x \in X_i$, $(i = 1, 2, \ldots, N)$.

If $p \geq 1$, then (2.1) becomes:

$$\left(\int_Y \left(\sum_{i=1}^N f_i(y)\right)^p v(y)\,d\nu(y)\right)^{\frac{1}{p}} \leq \sum_{i=1}^N \left(\int_Y f_i^p(y)v(y)\,d\nu(y)\right)^{\frac{1}{p}}, \tag{2.2}$$

i.e. the usual Minkowski inequality for integrals with N non-negative functions involved.

If $0 < p < 1$, then, with the obvious restrictions on the integrals, (2.2) holds in the reversed direction.

A standard proof of the Minkowski inequality (2.1) is just to use the corresponding Hölder inequality. In the continuous family situation, we even have the following remarkable equivalence theorem:

Theorem 2.3 *Let $f(x, y)$ be positive and measurable on $(X \times Y, \mu \times \nu)$. Assume that $p \geq 1$ and that $u(x)$ and $v(y)$ are weight functions. Then the following statements are equivalent:*

(i) *The continuous Hölder inequality* (1.2) *holds for all* X, $u(x)$ *and* $\mu(x)$ *such that* $\int_X u(x)\,d\mu(x) = 1$.
(ii) *The continuous Minkowski inequality* (2.1) *holds for all* X, $u(x)$ *and* $\mu(x)$ *such that* $\int_X u(x)\,d\mu(x) < \infty$.

As in the case of the Hölder inequality, we can also state the refinement of the continuous Minkowski inequality.

Theorem 2.4 *Let* $f(x, y)$ *be a non-negative and measurable function on* $(X \times Y, \mu \times \nu)$, *let* $u(x)$ *and* $v(y)$ *be weight functions on* X *and* Y, *respectively. Moreover, let* $\alpha(z, y)$ *be a non-negative function such that* $\int_Z \alpha(z, y)\,dz = 1$ *for* $y \in Y$. *If* $p \geq 1$, *then*

$$\int_Y \left(\int_X f(x, y) u(x)\,d\mu(x) \right)^p v(y)\,d\nu(y)$$
$$\leq \int_Z \left[\int_X \left(\int_Y \alpha(z, y) f^p(x, y) v(y)\,d\nu(y) \right)^{1/p} u(x)\,d\mu(x) \right]^p dz$$
$$\leq \left[\int_X \left(\int_Y f^p(x, y) v(y)\,d\nu(y) \right)^{1/p} u(x)\,d\mu(x) \right]^p .$$

3 Continuous/Family Popoviciu and Bellman-Type Inequalities

Popoviciu-type inequalities are some type of reversed Hölder-type inequalities. We just state one such result involving two functions:

Proposition 3.1 *Let* p, q *be positive real numbers such that* $\frac{1}{p} + \frac{1}{q} = 1$. *Let* f *and* g *be non-negative measurable functions on the measure space* (Y, ν). *Then the following inequality*

$$v_0 c_1 c_2 - \int_Y v(y) f(y) g(y)\,d\nu(y)$$
$$\geq \left(v_0 c_1^p - \int_Y v(y) f^p(y)\,d\nu(y) \right)^{\frac{1}{p}} \left(v_0 c_2^q - \int_Y v(y) g^q(y)\,d\nu(y) \right)^{\frac{1}{q}} \tag{3.1}$$

holds, where $v(y)$ *is a weight,*

$$v_0 c_1^p - \int_Y v(y) f^p(y)\,d\nu(y) \geq 0, \quad \text{and} \quad v_0 c_2^q - \int_Y v(y) g^q(y)\,d\nu(y) \geq 0.$$

Our continuous version of Proposition 3.1 reads:

Theorem 3.2 *Let $u(x)$ and $v(y)$ be weight functions on the measure spaces (X, μ) and (Y, ν), respectively, such that $\int_X u(x)\,d\mu(x) = 1$, let $f(x, y)$ be a positive measurable function on $X \times Y$, $v_0 \in (0, \infty)$, and assume that $f_0(x)$ is a function on X such that $v_0 f_0(x) > \int_Y f(x, y)v(y)\,d\nu(y)$, for all $x \in X$. Then the following continuous form of the Popoviciu inequality holds:*

$$\exp\left(\int_X \log(v_0 f_0(x))u(x)\,d\mu(x)\right)$$
$$-\int_Y \exp\left(\int_X \log f(x, y)u(x)\,d\mu(x)\right) v(y)\,d\nu(y)$$
$$\geq \exp\left[\int_X \log\left(v_0 f_0(x) - \int_Y f(x, y)v(y)\,d\nu(y)\right) u(x)\,d\mu(x)\right].$$

Example 3.3 Let $u(x) = 1$, $X = X_1 \cup X_2$, $X_1 \cap X_2 = \emptyset$ with $\int_{X_1} d\mu(x) = \frac{1}{p}$, $\int_{X_2} d\mu(x) = \frac{1}{q}$, where $\frac{1}{p} + \frac{1}{q} = 1$,

$$f_0(x) = \begin{cases} c_1^p, & x \in X_1 \\ c_2^q, & x \in X_2 \end{cases} \quad \text{and} \quad f(x, y) = \begin{cases} f^p(y), & x \in X_1 \\ g^q(y), & x \in X_2 \end{cases}.$$

Then we rediscover the Popoviciu inequality (3.1) in the finite form with $v_0 = 1$.

Theorem 3.2 also implies the following result of independent interest:

Corollary 3.4 *Let $w_1 > 0$, $w_2, \ldots, w_m \geq 0$ be reals, p, a_i, $i = 1, 2, \ldots, m$, be positive functions on X such that $\int_X \frac{d\mu(x)}{p(x)} = 1$ and a_i^p are measurable on X. Then*

$$w_1 \exp\left(\int_X \log a_1(x)\,d\mu(x)\right) - \sum_{i=2}^m w_i \exp\left(\int_X \log a_i(x)\,d\mu(x)\right)$$
$$\geq \exp\left\{\int_X \log\left[w_1(a_1(x))^{p(x)} - \sum_{i=2}^m w_i(a_i(x))^{p(x)}\right] \frac{d\mu(x)}{p(x)}\right\},$$

provided that all integrals exist.

Our refinement of the continuous Popoviciu inequality reads:

Theorem 3.5 *Let $u(x)$ and $v(y)$ be weight functions on the measure spaces (X, μ) and (Y, ν), respectively, where $\int_X u(x)\,d\mu(x) = 1$. Let $f(x, y)$ be a positive measurable function on $X \times Y$, $v_0 \geq 0$, and assume that $f_0(x)$ is a function on X such that $v_0 f_0(x) > \int_Y f(x, y)v(y)\,d\nu(y)$, for all $x \in X$. Moreover, let $\alpha(z, y)$*

be a non-negative integrable function on $Z \times Y$ such that $\int_Z \alpha(z, y)\, dz = 1$ for $y \in Y$, where (Z, dz) is a σ–finite measure space. Then the following refinement of the continuous form of the Popoviciu inequality holds:

$$\begin{aligned}
&\exp\left(\int_X \log\left(v_0 f_0(x)\right) u(x)\, d\mu(x)\right) \\
&\quad - \int_Y \exp\left(\int_X \log(f(x, y)) u(x)\, d\mu(x)\right) v(y)\, d\nu(y) \\
&\quad \geq \exp\left(\int_X \log\left(v_0 f_0(x)\right) u(x)\, d\mu(x)\right) \\
&\quad - \int_Z \left[\exp \int_X \log\left(\int_Y f(x, y)\alpha(z, y) v(y)\, d\nu(y)\right) u(x)\, d\mu(x)\right] dz \\
&\quad \geq \exp\left[\int_X \log\left(v_0 f_0(x) - \int_Y f(x, y) v(y)\, d\nu(y)\right) u(x)\, d\mu(x)\right] \geq 0.
\end{aligned}$$

Example 3.6 Using a suitable choice for the measure μ and functions u and f, we get the following refinement of the integral Popoviciu inequality (3.1) for two functions with $v_0 = 1$.

$$\begin{aligned}
&c_1 c_2 - \int_Y v(y) f(y) g(y)\, d\nu(y) \\
&\geq c_1 c_2 - \int_Z \left(\int_Y \alpha(z, y) f^p(y) v(y)\, d\nu(y)\right)^{1/p} \\
&\left(\int_Y \alpha(z, y) g^q(y) v(y)\, d\nu(y)\right)^{1/q} dz \\
&\geq \left(c_1^p - \int_Y v(y) f^p(y)\, d\nu(y)\right)^{1/p} \left(c_2^q - \int_Y v(y) g^q(y)\, d\nu(y)\right)^{1/q}.
\end{aligned}$$

Bellman-type inequalities are just a type of reversed Minkowski-type inequalities in a similar way as the Popoviciu inequality was related to the Hölder inequality. Our continuous form of this inequality reads:

Theorem 3.7 *Let $f_0(x)$, $f(x, y)$, v_0, $u(x)$, $v(y)$, X, Y, μ, ν be defined as in Theorem 3.2. Then, for $p \geq 1$,*

$$\begin{aligned}
&\left(\int_X \left[v_0 f_0^p(x) - \int_Y f^p(x, y) v(y)\, d\nu(y)\right]^{\frac{1}{p}} u(x)\, d\mu(x)\right)^p \\
&\qquad \leq v_0 \left[\int_X f_0(x) u(x)\, d\mu(x)\right]^p - \int_Y \left[\int_X f(x, y) u(x)\, d\mu(x)\right]^p v(y)\, d\nu(y),
\end{aligned}$$

whenever $v_0 f_0^p(x) \geq \int_Y f^p(x, y) v(y)\, d\nu(y)$, *for all* $x \in X$.

Example 3.8 By applying Theorem 3.7 with $u(x) = v(y) = 1$, $v_0 = 1$, $X = \bigcup_{i=1}^n X_i$, $X_i = [i-1, i)$ for $i = 1, 2, \ldots, n$, $d\mu(x) = dx$, $f(x, y) = f_i(y)$ and $f_0(x) = c_i$ for each $x \in X_i$, $i = 1, 2, \ldots, n$, we get the following version of Bellman's inequality:

$$\sum_{i=1}^n \left(c_i^p - \int_Y f_i^p(y) d\nu(y)\right)^{\frac{1}{p}} \leq \left(\left(\sum_{i=1}^n c_i\right)^p - \int_Y \left(\sum_{i=1}^n f_i(y)\right)^p d\nu(y)\right)^{\frac{1}{p}}$$

whenever

$$c_i \geq \left(\int_Y f_i^p(y)\, d\nu(y)\right)^{\frac{1}{p}}, \quad i = 1, 2, \ldots, n.$$

Our refinement of the continuous Bellman inequality reads:

Theorem 3.9 *Let the assumptions of Theorem 2.4 hold. Assume that* $v_0 \geq 0$ *and* $f_0(x)$ *is a function on* X *such that* $v_0 f_0^p(x) - \int_Y f^p(x, y)\, v(y)\, d\nu(y) \geq 0$. *Let* $\alpha(z, y)$ *satisfies the assumptions of Theorem 3.5.*

Then the following refinement of the continuous Bellman inequality holds for $p \geq 1$:

$$\begin{aligned} v_0 &\left[\int_X f_0(x) u(x)\, d\mu(x)\right]^p - \int_Y \left(\int_X f(x, y) u(x)\, d\mu(x)\right)^p v(y)\, d\nu(y) \\ &\geq v_0 \left[\int_X f_0(x) u(x)\, d\mu(x)\right]^p \\ &\quad - \int_Z \left[\int_X \left(\int_Y \alpha(z, y) f^p(x, y) v(y)\, d\nu(y)\right)^{1/p} u(x)\, d\mu(x)\right]^p dz \\ &\geq \left(\int_X \left[v_0 f_0^p(x) - \int_Y f^p(x, y) v(y)\, d\nu(y)\right]^{\frac{1}{p}} u(x)\, d\mu(x)\right)^p. \end{aligned}$$

4 Continuous Beckenbach-Dresher-Type Inequalities

The Beckenbach-Dresher inequality in its original form looks like a Minkowski-type inequality with the Gini means involved. Inspired by an extension of this result by Peetre-Persson, the continuous version reads:

Theorem 4.1 *Let (X, μ), (Y, ν) and (Y, λ) be measure spaces. Let f, g be non-negative functions on $X \times Y$ such that f is integrable with respect to the measure $\mu \times \nu$ and g is integrable with respect to $\mu \times \lambda$.*

(a) If

(i) $u \geq 1$ and $q \leq 1 \leq p$ $(q \neq 0)$, or
(ii) $u < 0$ and $p \leq 1 \leq q$ $(p \neq 0)$, and all terms exist, then

$$\frac{\left(\int_Y \left(\int_X f(x,y)d\mu(x)\right)^p d\nu(y)\right)^{\frac{u}{p}}}{\left(\int_Y \left(\int_X g(x,y)d\mu(x)\right)^q d\lambda(y)\right)^{\frac{u-1}{q}}} \leq \int_X \frac{\left(\int_Y f^p(x,y)d\nu(y)\right)^{\frac{u}{p}}}{\left(\int_Y g^q(x,y)d\lambda(y)\right)^{\frac{u-1}{q}}} d\mu(x). \tag{4.1}$$

If (iii) $0 < u \leq 1$, $p \leq 1$ and $q \leq 1$, $p, q \neq 0$, then inequality (4.1) *is reversed.*

(b) If $u \geq 1$ and $p \geq 1$, then

$$\left(\int_Y \left(\int_X f(x,y)d\mu(x)\right)^p d\nu(y)\right)^{\frac{u}{p}} \times \exp\left(\frac{1-u}{\int_Y d\lambda}\int_Y \log\left(\int_X g(x,y)d\mu(x)\right) d\lambda(y)\right)$$
$$\leq \int_X \left(\int_Y f^p(x,y)d\nu(y)\right)^{\frac{u}{p}} \exp\left(\frac{1-u}{\int_Y d\lambda}\int_Y \log g(x,y)d\lambda(y)\right) d\mu(x).$$

Remark 4.2 Using a discrete measure μ in (4.1), we can formulate the following Beckenbach-Dresher inequality (of Peetre-Persson type) for integrals: If u, p, q satisfy the assumptions of Theorem 4.1 (a), then

$$\frac{\left(\int_Y \left(\sum_{i=1}^n f_i(y)\right)^p d\nu(y)\right)^{\frac{u}{p}}}{\left(\int_Y \left(\sum_{i=1}^n g_i(y)\right)^q d\lambda(y)\right)^{\frac{u-1}{q}}} \leq \sum_{i=1}^n \frac{\left(\int_Y f_i^p(y)d\nu(y)\right)^{\frac{u}{p}}}{\left(\int_Y g_i^q(y)d\lambda(y)\right)^{\frac{u-1}{q}}}.$$

5 Final Remarks

Proofs of all results in this paper can be found in papers [1–5] and in the book [6]. In this book, we also pointed out that the close connection between convexity and interpolation theory is partly kept also in this general continuous/family situation. In particular, in the Appendix in [6], it is given a short introduction to the theory of interpolation between infinite many Banach spaces, even between families of Banach spaces (in the classical case only two or finite many Banach spaces are involved). It was in this connection two of the present authors first met the need to prove a classical inequality of Hölder-type in continuous/family form. We also mention that this new book manuscript contains more than 100 references on the subject, several other examples of classical inequalities in continuous/family form (e.g. those by Jensen, Jensen-Mercer and Hardy) and a view on some inequalities in Banach lattice norm settings. Finally, we pronounce that the research presented in this paper implies, in our opinion, an interesting direction of research, namely to prove some corresponding continuous/family forms of other inequalities than those covered in the book [6].

References

1. L.I. Nikolova, L.E. Persson, S. Varošanec, Continuous forms of classical inequalities. Mediterr. J. Math. **13**, 3483–3497 (2016)
2. L.I. Nikolova, L.E. Persson, S. Varošanec, Refinement of continuous forms of classical inequalities. Eurasian Math. J. **12**(2), 59–73 (2021)
3. L.I. Nikolova, L.E. Persson, S. Varošanec, Continuous refinements of some Jensen-type inequalities via strong convexity with applications. J. Inequal. Appl. **2022**, Paper No. 63, 15 pp. (2022)
4. L.I. Nikolova, L.E. Persson, S. Varošanec, M.F. Yimer, Refinements of some classical inequalities via superquadraticity. J. Inequal. Appl. **2022**, Paper No. 86, 15 pp. (2022)
5. L. I. Nikolova, S. Varošanec, Continuous forms of Gauss-Pólya type inequalities involving derivatives. Math. Inequal. Appl. **22**(4), 1385–1395 (2019)
6. L. Nikolova, L. E. Persson and S. Varošanec, Continuous Versions of Some Classical Inequalities and Their Refinements, Book manuscript, 2023, submitted.

Approximation by Vilenkin-Nörlund Means in Lebesgue Spaces

Nino Anakidze, Nika Areshidze, and Lasha Baramidze

Abstract In this paper we improve and complement a result by Móricz and Siddiqi (J Approx Theory 70(3):375–389, 1992). In particular, we prove that their estimate of the Nörlund means with respect to the Vilenkin system holds also without their additional condition. Moreover, we prove a similar approximation result in Lebesgue spaces for any $1 \leq p < \infty$.

Keywords Vilenkin group · Vilenkin system · Fejér means · Nörlund means · Approximation

2010 Mathematics Subject Classification 42C10, 42B30

1 Introduction

Concerning some definitions and notations used in this introduction we refer to Sect. 2.

It is well-known (see e.g. [20, 39, 56]) that, for any $1 \leq p \leq \infty$ and $f \in L_p(G_m)$, there exists an absolute constant C_p, depending only on p such that

$$\|\sigma_n f\|_p \leq C_p \|f\|_p .$$

The research was supported by Shota Rustaveli National Science Foundation (SRNSF) grant no. MR-23-286.

N. Anakidze
School of Science and Technology, The University of Georgia, Tbilisi, Georgia
e-mail: nino.anakidze@ug.edu.ge

N. Areshidze · L. Baramidze (✉)
Faculty of Exact and Natural Sciences, Department of Mathematics, Tbilisi State University, Tbilisi, Georgia

R. Duduchava et al. (eds.), *Tbilisi Analysis and PDE Seminar*, Research Perspectives Ghent Analysis and PDE Center 7, https://doi.org/10.1007/978-3-031-62894-8_2

Moreover, (for details see [39]) if $1 \leq p \leq \infty$, $M_N \leq n < M_{N+1}$, $f \in L^p(G_m)$ and $n \in \mathbb{N}$, then

$$\|\sigma_n f - f\|_p \leq 2R^5 \sum_{s=0}^{N} \frac{M_s}{M_N} \omega_p \left(1/M_s, f\right), \tag{1.1}$$

where $R := \sup_{k\in\mathbb{N}} m_k$ and $\omega_p(\delta, f)$ is the modulus of continuity of L^p, $1 \leq p \leq \infty$ functions defined by

$$\omega_p(\delta, f) = \sup_{|t|<\delta} \|f(x+t) - f(x)\|, \qquad \delta > 0.$$

It follows that if $f \in lip\,(\alpha, p)$, i.e.,

$$lip(\alpha, p) := \{f \in L^p : \omega_p(\delta, f) = O(\delta^\alpha) \quad \text{as} \quad \delta \to 0\},$$

then

$$\|\sigma_n f - f\|_p = \begin{cases} O\left(1/M_N\right), & \text{if} \quad \alpha > 1, \\ O\left(N/M_N\right), & \text{if} \quad \alpha = 1, \\ O\left(1/M_n^\alpha\right), & \text{if}\, \alpha < 1. \end{cases}$$

Moreover, (for details see [39]) if $1 \leq p < \infty$, $f \in L^p(G)$ and

$$\left\|\sigma_{M_n} f - f\right\|_p = o\left(1/M_n\right), \text{ as } n \to \infty,$$

then f is a constant function.

The weak-(1, 1) type inequality for the maximal operators of Vilenkin-Fejer means σ^*, defined by

$$\sigma^* f = \sup_{n\in\mathbb{N}} |\sigma_n f|$$

can be found in Schipp [40] for Walsh series and in PÍ, Simon [33] and Weisz [54] for bounded Vilenkin series. Boundedness of the maximal operators of Vilenkin-Féjer means of the one- and two-dimensional cases can be found in Fridli [15], Gát [17], Goginava [18], Nagy and Tephnadze [29–32], Simon [42, 43], Tephnadze [44–49], Tutberidze [37, 50], Weisz [25, 38, 55].

Convergence and summability of Nörlund means with respect to Vilenkin systems were studied by Areshidze and Tephnadze [3] (see also [2]), Baramidze, Persson and Tephnadze [4–8], Blahota et al. [13] (see also [10–12, 34–36]), Fridli et al. [16], Goginava [19], Nagy [26–28] (see also [9, 14]), Memic [21, 22].

Móricz and Siddiqi [24] investigated the approximation properties of some special Nörlund means of Walsh-Fourier series of L_p functions in norm. In particular, they proved that if $f \in L_p(G)$, $1 \leq p \leq \infty$, $n = 2^j + k$, $1 \leq k \leq 2^j$ $(n \in \mathbb{N}_+)$ and $(q_k, k \in \mathbb{N})$ is a sequence of non-negative numbers, such that

$$\frac{n^{\gamma-1}}{Q_n^\gamma} \sum_{k=0}^{n-1} q_k^\gamma = O(1), \quad \text{for some } 1 < \gamma \leq 2, \tag{1.2}$$

then, there exists an absolute constant C_p, depending only on p such that

$$\|t_n f - f\|_p \leq \frac{C_p}{Q_n} \sum_{i=0}^{j-1} 2^i q_{n-2^i} \omega_p \left(\frac{1}{2^i}, f \right) + C_p \omega_p \left(\frac{1}{2^j}, f \right), \tag{1.3}$$

when the sequence $(q_k, k \in \mathbb{N})$ is non-decreasing, while

$$\|t_n f - f\|_p \leq \frac{C_p}{Q_n} \sum_{i=0}^{j-1} \left(Q_{n-2^i+1} - Q_{n-2^{i+1}+1} \right) \omega_p \left(\frac{1}{2^i}, f \right) + C_p \omega_p \left(\frac{1}{2^j}, f \right),$$

when the sequence $(q_k, k \in \mathbb{N})$ is non-increasing.

In this paper we improve and complement a result by Móricz and Siddiqi [24]. In particular, we prove that their estimate of the Nörlund means with respect to the Vilenkin system holds also without their additional condition. Moreover, we prove a similar approximation result in Lebesgue spaces for any $1 \leq p < \infty$.

2 Preliminaries

Let $\mathbb{N}_+$ denote the set of the positive integers, $\mathbb{N} := \mathbb{N}_+ \cup \{0\}$. Let $m =: (m_0, m_1, \ldots)$ be a sequence of positive integers not less than 2. Denote by

$$Z_{m_k} := \{0, 1, \ldots, m_k - 1\}$$

the additive group of integers modulo m_k. Define the group G_m as the complete direct product of the group Z_{m_k} with the product of the discrete topologies of Z_{m_k}'s.

The direct product μ of the measures

$$\mu_k (\{j\}) := 1/m_k \quad (j \in Z_{m_k})$$

is the Haar measure on G_m with $\mu (G_m) = 1$.

If $\sup_{k\in\mathbb{N}} m_k < +\infty$, then we call G_m a bounded Vilenkin group. If $\{m_k\}_{k\geq 0}$ sequence is unbounded, then G_m is said to be unbounded Vilenkin group. In this paper we consider only bounded Vilenkin groups.

The elements of G_m are represented by the sequences

$$x := (x_0, x_1, \ldots, x_k, \ldots) \qquad \left(x_k \in Z_{m_k}\right).$$

It is easy to give a base for the neighborhood of G_m, namely

$$I_0(x) := G_m, \quad I_n(x) := \{y \in G_m \mid y_0 = x_0, \ldots, y_{n-1} = x_{n-1}\} \; (x \in G_m, \; n \in \mathbb{N}).$$

For the simplicity we also define by I_n as $I_n := I_n(0)$.

Let us define a generalized number system based on m in the following way:

$$M_0 =: 1, \quad M_{k+1} =: m_k M_k \quad (k \in \mathbb{N})$$

Then every $n \in \mathbb{N}$ can be uniquely expressed as

$$n = \sum_{k=0}^{\infty} n_j M_j, \quad \text{where} \quad n_j \in Z_{m_j} \quad (j \in \mathbb{N})$$

and only a finite number of n_j's differ from zero.

In 1947 Vilenkin [51–53] investigated a group G_m and introduced the Vilenkin systems $\{\psi_j\}_{j=0}^{\infty}$ as

$$\psi_n(x) := \prod_{k=0}^{\infty} r_k^{n_k}(x) \qquad (n \in \mathbb{N}).$$

where $r_k(x)$ are the generalized Rademacher functions defined by

$$r_k(x) := \exp(2\pi i x_k / m_k), \quad (k \in \mathbb{N}).$$

These systems include as a special case the Walsh system when $m_k = 2$ for any $k \in \mathbb{N}$.

The norms (or quasi-norms) of Lebesgue spaces $L_p(G_m)$ are defined by

$$\|f\|_p^p := \int_{G_m} |f|^p \, d\mu.$$

The Vilenkin system is orthonormal and complete in $L^2(G_m)$ (for details see e.g. [1] and [41]).

If $f \in L^1(G_m)$, we can define the Fourier coefficients, partial sums of the Fourier series and Fejér means with respect to the Vilenkin system as

$$\widehat{f}(k) := \int_{G_m} f\overline{\psi}_k d\mu, \quad (k \in \mathbb{N}),$$

$$S_n f := \sum_{k=0}^{n-1} \widehat{f}(k)\psi_k, \quad (n \in \mathbb{N}_+, \ S_0 f := 0),$$

$$\sigma_n f := \frac{1}{n}\sum_{k=1}^{n} S_k f, \quad (n \in \mathbb{N}_+).$$

The n-th Nörlund mean t_n of the Vilenkin-Fourier series of a integrable function f is defined by

$$t_n f := \frac{1}{Q_n}\sum_{k=1}^{n} q_{n-k} S_k f, \quad \text{where} \quad Q_n := \sum_{k=0}^{n-1} q_k. \tag{2.1}$$

Here $\{q_k : k \geq 0\}$ is a sequence of nonnegative numbers, where $q_0 > 0$ and

$$\lim_{n\to\infty} Q_n = \infty.$$

Then the summability method (2.1) generated by $\{q_k : k \geq 0\}$ is regular if and only if (see [23])

$$\lim_{n\to\infty} \frac{q_{n-1}}{Q_n} = 0.$$

In this paper we investigate regular Nörlund means only.

It is well-known (for details see e.g. [39]) that every Nörlund summability method generated by non-increasing sequence $(q_k, k \in \mathbb{N})$ is regular, but Nörlund means generated by non-decreasing sequence $(q_k, k \in \mathbb{N})$ is not always regular.

The representation

$$t_n f(x) = \int_{G_m} f(t) F_n(x-t) d\mu(t)$$

play central roles in the sequel, where

$$F_n =: \frac{1}{Q_n}\sum_{k=1}^{n} q_{n-k} D_k \tag{2.2}$$

is called the kernels of the Nörlund means.

3 Formulation of Main Results

Based on estimate (1.1) we can prove our next main results:

Theorem 3.1 *Let $M_N \leq n < M_{N+1}$ and t_n be a regular Nörlund mean generated by non-decreasing sequence $\{q_k : k \in \mathbb{N}\}$, in sign $q_k \uparrow$. Then, for some $f \in L^p(G_m)$, where $1 \leq p < \infty$,*

$$\|t_n f - f\|_p \leq \frac{3R^3}{Q_n}\sum_{i=0}^{N-1} M_i q_{n-M_i}\omega_p\left(\frac{1}{M_i}, f\right) + 2R^3\omega_p\left(\frac{1}{M_N}, f\right).$$

Theorem 3.2 *Let t_n be Nörlund mean generated by non-increasing sequence $\{q_k : k \in \mathbb{N}\}$, in sign $q_k \downarrow$. Then, for some $f \in L^p(G_m)$, where $1 \leq p < \infty$,*

$$\|t_{M_n} f - f\|_p \leq 3R^2\sum_{s=0}^{n}\frac{M_s}{M_n}\omega_p\left(1/M_s, f\right) + C\sum_{s=0}^{n-1}\frac{(n-s)M_s}{M_n}\frac{q_{M_s}}{q_{M_n}}\omega_p\left(1/M_s, f\right).$$

Theorem 3.3 *Let $M_N \leq n < M_{N+1}$ and t_n be Nörlund mean generated by non-increasing sequence $\{q_k : k \in \mathbb{N}\}$, in sign $q_k \downarrow$, satisfying the condition*

$$\frac{1}{Q_n} = O\left(\frac{1}{n}\right), \quad as \quad n \to \infty. \tag{3.1}$$

Then, for some $f \in L^p(G_m)$, where $1 \leq p < \infty$,

$$\|t_n f - f\|_p \leq C\sum_{j=0}^{N}\frac{M_j}{M_N}\omega_p\left(1/M_j, f\right).$$

As a consequence we obtain the following similar result proved in [24]:

Corollary 3.4 *Let $\{q_k : k \geq 0\}$ be a sequence of non-negative numbers such that in case $q_k \uparrow$ condition*

$$\frac{q_{n-1}}{Q_n} = O\left(\frac{1}{n}\right), \quad as \quad n \to \infty. \tag{3.2}$$

is satisfied, while in case $q_k \downarrow$ condition (3.1) is satisfied. If $f \in Lip(\alpha, p)$ for some $\alpha > 0$ and $1 \leq p < \infty$, then

$$\|t_n f - f\|_p = \begin{cases} O(n^{-\alpha}), & if \quad 0 < \alpha < 1, \\ O(n^{-1}\log n), & if \quad \alpha = 1, \\ O(n^{-1}), & if \quad \alpha > 1, \end{cases} \tag{3.3}$$

As a consequence we obtain the following similar result proved in Móricz and Siddiqi [24]:

Corollary 3.5

(a) *Let t_n be Nörlund means generated by non-decreasing sequence $\{q_k : k \in \mathbb{N}\}$ satisfying regularity condition* (3.2). *If $f \in Lip(\alpha, p)$ for some $\alpha > 0$ and $1 \le p < \infty$, then $t_n f$ converge to f in $L_p(G_m)$ norm.*

(b) *Let t_n be Nörlund mean generated by non-increasing sequence $\{q_k : k \in \mathbb{N}\}$ satisfying condition* (3.1). *If $f \in Lip(\alpha, p)$ for some $\alpha > 0$ and $1 \le p < \infty$, then $t_n f$ converge to f in $L_p(G_m)$ norm.*

References

1. G.N. Agaev, N.Ya. Vilenkin, G.M. Dzhafarly, A.I. Rubinshtein, *Multiplicative Systems of Functions and Harmonic Analysis on Zero-dimensional Groups* (Baku, Ehim, 1981) (in Russian)
2. N. Anakidze, N. Areshidze, L. Baramidze, Approximation by Nörlund means with respect to Vilenkin system in Lebesgue spaces. Acta Math. Hungar. **172**, 529–542 (2024)
3. N. Areshidze, G. Tephnadze, Approximation by Nörlund means with respect to Walsh system in Lebesgue spaces. Math. Inequal. Appl. **27**(1), 137–147 (2024)
4. L. Baramidze, L.E. Persson, G. Tephnadze, P. Wall, Sharp $H_p - L_p$ type inequalities of weighted maximal operators of Vilenkin-Nörlund means and its applications. J. Inequal. Appl. **2016**, paper no. 242, 20 pp. (2016)
5. D. Baramidze, L.-E. Persson, H. Singh, G. Tephnadze, Some new results and inequalities for subsequences of Nörlund logarithmic means of Walsh-Fourier series. J. Inequal. Appl. **2022**, paper no. 30, 13 pp. (2022)
6. D. Baramidze, L. Baramidze, L.-E. Persson, G. Tephnadze, Some now restricted maximal operators of Fejér means of Walsh-Fourier series. Banach J. Math. Anal. **17**(4), 75 (2023)
7. D. Baramidze, N. Nadirashvili, L.-E. Persson, G. Tephnadze, Some weak-type inequalities and almost everywhere convergence of Vilenkin-Nörlund means. J. Inequal. Appl. **2023**, paper no. 66, 17 pp. (2023)
8. D. Baramidze, L.-E. Persson, K. Tangrand, G. Tephnadze, $(H_p - L_p)$ type inequalities for subsequences of Nörlund means of Walsh-Fourier series. J. Inequal. Appl., paper no. 52, 13 pp. (2023)
9. I. Blahota, K. Nagy, Approximation by Θ-means of Walsh-Fourier series. Anal. Math. **44**(1), 57–71 (2018)
10. I. Blahota, G. Tephnadze, On the (C, α)-means with respect to the Walsh system. Anal. Math. **40**, 161–174 (2014)
11. I. Blahota, G. Tephnadze, A note on maximal operators of Vilenkin-Nörlund means. Acta Math. Acad. Paed. Nyireg. **32**, 203–213 (2016)
12. I. Blahota, G. Tephnadze, R. Toledo, Strong convergence theorem of (C, α)-means with respect to the Walsh system. Tohoku Math. J. **67**(4), 573–584 (2015)
13. I. Blahota, L.E. Persson, G. Tephnadze, On the Nörlund means of Vilenkin-Fourier series. Czech. Math J. **65**(4), 983–1002 (2015)
14. I. Blahota, K. Nagy, G. Tephnadze, Approximation by Marcinkiewicz Θ-means of double Walsh-Fourier series. Math. Inequal. Appl. **22**(3), 837–853 (2019)
15. S. Fridli, On the rate of convergence of Cesaro means of Walsh-Fourier series. J. Approx. Theory **76**(1), 31–53 (1994)

16. S. Fridli, P. Manchanda, A. Siddiqi, Approximation by Walsh-Nörlund means. Acta Sci. Math. **74**(3–4), 593–608 (2008)
17. G. Gát, Cesàro means of integrable functions with respect to unbounded Vilenkin systems. J. Approx. Theory **124**(1), 25–43 (2003)
18. U. Goginava, Marcinkiewicz-Fejer means of d-dimensional Walsh-Fourier series. J. Math. Anal. Appl. **307**(1), 206–218 (2015)
19. U. Goginava, Maximal operators of Walsh-Nörlund means on the dyadic Hardy spaces. Acta Math. Hungar. **169**(1), 171–190 (2023)
20. B.I. Golubov, A.V. Efimov, V.A. Skvortsov, *Walsh Series and Transforms* (Kluwer Academic Publishers Group, Dordrecht, 1991)
21. N. Memić, An estimate of the maximal operator of the Nörlund logarithmic means with respect to the character system of the group of 2-adic integers on the Hardy space H_1. Bull. Iranian Math. Soc. **48**(6), 3381–3391 (2022)
22. N. Memić, L.E. Persson, G. Tephnadze, A note on the maximal operators of Vilenkin-Nörlund means with non-increasing coefficients. Stud. Sci. Math. Hung. **53**(4), 545–556 (2016)
23. C.N. Moore, *Summable Series and Convergence Factors*, Summable Series and Convergence Factors (Dover Publications, New York, 1966)
24. F. Móricz, A. Siddiqi, Approximation by Nörlund means of Walsh-Fourier series. J. Approx. Theory **70**(3), 375–389 (1992)
25. N. Nadirashvili, L.-E. Persson, G. Tephnadze, F. Weisz, Vilenkin-Lebesgue points and almost everywhere convergence of Vilenkin-Fejér means and applications. Mediterr. J. Math. **19**(5), 239 (2022)
26. K. Nagy, Approximation by Nörlund means of quadratical partial sums of double Walsh-Fourier series. Anal. Math. **36**(4), 299–319 (2010)
27. K. Nagy, Approximation by Nörlund means of Walsh-Kaczmarz-Fourier series. Georgian Math. J. **18**(1), 147–162 (2011)
28. K. Nagy, Approximation by Nörlund means of double Walsh-Fourier series for Lipschitz functions. Math. Inequal. Appl. **15**(2), 301–322 (2012)
29. K. Nagy, G. Tephnadze, Walsh-Marcinkiewicz means and Hardy spaces. Cent. Eur. J. Math. **12**(8), 1214–1228 (2014)
30. K. Nagy, G. Tephnadze, Approximation by Walsh-Marcinkiewicz means on the Hardy space. Kyoto J. Math. **54**(3), 641–652 (2014)
31. K. Nagy, G. Tephnadze, Strong convergence theorem for Walsh-Marcinkiewicz means. Math. Inequal. Appl. **19**(1), 185–195 (2016)
32. K. Nagy, G. Tephnadze, The Walsh-Kaczmarz-Marcinkiewicz means and Hardy spaces. Acta Math. Hung. **149**(2), 346–374 (2016)
33. J. Pál, P. Simon, On a generalization of the concept of derivative. Acta Math. Hung. **29**, 155–164 (1977)
34. L.E. Persson, G. Tephnadze, P. Wall, On the maximal operators of Vilenkin-Nörlund means. J. Fourier Anal. Appl. **21**(1), 76–94 (2015)
35. L.E. Persson, G. Tephnadze, P. Wall, Some new (H_p, L_p) type inequalities of maximal operators of Vilenkin-Nörlund means with non-decreasing coefficients. J. Math. Inequal. **9**(4), 1055–1069 (2015)
36. L.E. Persson, G. Tephnadze, P. Wall, On the Nörlund logarithmic means with respect to Vilenkin system in the martingale Hardy space H_1. Acta Math. Hung. **154**(2), 289–301 (2018)
37. L.E. Persson, G. Tephnadze, G. Tutberidze, On the boundedness of subsequences of Vilenkin-Fejér means on the martingale Hardy spaces. Oper. Matrices **14**(1), 283–294 (2020)
38. L.-E. Persson, F. Schipp, G. Tephnadze, F. Weisz, An analogy of the Carleson-Hunt theorem with respect to Vilenkin systems. J. Fourier Anal. Appl. **28**, 48, 1–29 (2022)
39. L.E. Persson, G. Tephnadze, F. Weisz, *Martingale Hardy Spaces and Summability of One-dimensional Vilenkin-Fourier Series* (Birkhäuser/Springer, 2022), book manuscript
40. F. Schipp, Certain rearrangements of series in the Walsh series. Mat. Zametki **18**, 193–201 (1975)

41. F. Schipp, W.R. Wade, P. Simon, J. Pál, *Walsh Series. An Introduction to Dyadic Harmonic Analysis* (Adam Hilger, Bristol, 1990)
42. P. Simon, Strong convergence of certain means with respect to the Walsh-Fourier series. Acta Math. Hungar. **49**(3–4), 425–431 (1987)
43. P. Simon, On the Cesáro summability with respect to the Walsh–Kaczmarz system. J. Approx. Theory **106**(2), 249–261 (2000)
44. G. Tephnadze, Fejér means of Vilenkin-Fourier series. Stud. Sci. Math. Hung. **49**(1), 79–90 (2012)
45. G. Tephnadze, On the maximal operators of Vilenkin-Fejér means on Hardy spaces. Math. Inequal. Appl. **16**(2), 301–312 (2013)
46. G. Tephnadze, On the maximal operators of Vilenkin-Fejér means. Turk. J. Math. **37**, 308–318 (2013)
47. G. Tephnadze, On the maximal operators of Walsh-Kaczmarz-Fejér means. Period. Math. Hung. **67**(1), 33–45 (2013)
48. G. Tephnadze, Approximation by Walsh-Kaczmarz-Fejér means on the Hardy space. Acta Math. Sci. **34**(5), 1593–1602 (2014)
49. G. Tephnadze, The one-dimensional martingale hardy spaces and partial sums and Fejér means with respect to Walsh system. Mem. Differential Equations Math. Phys. **88**, 109–158 (2023)
50. G. Tutberidze, Modulus of continuity and boundedness of subsequences of Vilenkin-Fejer means in the martingale Hardy spaces. Geo. Math. J. **29**(1), 153–162 (2022)
51. N.Y. Vilenkin, On the theory of lacunary orthogonal systems (Russian). Izv. Akad. Nauk SSSR Ser. Mat. **13**, 245–252 (1949)
52. N.Y. Vilenkin, On the theory of Fourier integrals on topological groups. Mat. Sbornik N. S. **30**(72), 233–244 (1952)
53. N.Y. Vilenkin, On a class of complete orthonormal systems. Amer. Math. Soc. Trans. **28**(2), 1–35 (1963)
54. F. Weisz, *Martingale Hardy Spaces and their Applications in Fourier Analysis* (Springer, Berlin, 1994)
55. F. Weisz, Hardy spaces and Cesàro means of two-dimensional Fourier series. Bolyai Soc. Math. Stud. **5**, 353–367 (1996)
56. A. Zygmund, *Trigonometric Series*, vol. 1 (Cambridge University Press, 1959)

On Divergence of Fejér Means with Respect to Walsh System on Sets of Measure Zero

Nika Areshidze, Lars-Erik Persson, and George Tephnadze

Abstract In this paper we prove by the concrete construction that for any set E of measure zero there exists a function $f \in L_p(G)(1 \leq p < \infty)$ such that the Féjer means with respect to Wals system diverge on this set. The key is to use new constructions of Walsh polynomials, which was introduced in Persson et al. (Martingale Hardy Spaces and Summability of One-dimensional Vilenkin-Fourier Series. Birkhäuser/Springer, 2022). In fact, the theorem we prove follows from the general result of Karagulyan (Mat. Sb. 202(1):11–36, 2011), but we provide an alternative approach and the constructed function in our proof has a simpler explicit representation.

Keywords Fourier analysis · Walsh system · Walsh group · Walsh-Fourier series · Almost everywhere convergence · Kolmogorov theorem · Divergence of set of measure zero

2000 Mathematics Subject Classification 42C10, 42B25

The research was supported by Shota Rustaveli National Science Foundation grant no. MR-23-173.

N. Areshidze (✉)
Department of Mathematics, Faculty of Exact and Natural Sciences, Tbilisi State University, Tbilisi, Georgia

L.-E. Persson
UiT The Arctic University of Norway, Narvik, Norway

Department of Mathematics and Computer Science, Karlstad University, Karlstad, Sweden

G. Tephnadze
School of Science and Technology, The University of Georgia, Tbilisi, Georgia
e-mail: g.tephnadze@ug.edu.ge

R. Duduchava et al. (eds.), *Tbilisi Analysis and PDE Seminar*,
Research Perspectives Ghent Analysis and PDE Center 7,
https://doi.org/10.1007/978-3-031-62894-8_3

1 Introduction Including Some Historical Results

The almost-everywhere convergence of Fourier series for L_2 functions was postulated by Luzin [24] in 1915 and the problem was known as Luzin's conjecture. Carleson's theorem is a fundamental result in mathematical analysis establishing the pointwise (Lebesgue) almost everywhere convergence of Fourier series of L_2 functions, proved by Carleson [9] in 1966. The name is also often used to refer to the extension of the result by Hunt [15], which was given in 1968 to L_p functions for $p \in (1, \infty)$ (also known as the Carleson-Hunt theorem). Carleson's original proof is exceptionally hard to read, and although several authors have simplified the arguments there are still no easy proofs of his theorem. Fefferman [11] published a new proof of Hunt's extension, which was done by bounding a maximal operator of partial sums $S^*f := \sup_{n\in\mathbb{N}} |S_n f|$. This, in its turn, inspired to derive a much simplified proof of the L_2 result by Lacey and Thiele [23], explained in more detail in Lacey [22]. In the books Fremlin [12] and Grafakos [14] were also given proofs of Carleson's theorem. Already in 1923, Kolmogorov [21] showed that the analogue of Carleson's result for L_1 is false by finding such a function whose Fourier series diverges almost everywhere (improved slightly in 1926 to diverging everywhere).

Almost everywhere convergence of subsequences of partial sums with respect to Walsh system in the Lebesgue space $L_1(G)$ can be found in [2]. The analogue of Carleson's theorem for the Walsh system was proved by Billard [4] for $p = 2$ and by Sjölin [34] for $1 < p < \infty$. Schipp [32] investigated the so called tree martingales generalized the results about the maximal function, quadratic variation and martingale transforms to these martingales. Using these results, he gave a proof of Carleson's theorem for Walsh-Fourier series. A proof of almost everywhere convergence of Walsh-Fourier series was also given by Demeter [10] in 2015. Stein [36] constructed an integrable function whose Walsh-Fourier series diverges almost everywhere. Later Schipp [30] (see also [33]) proved that there exists an integrable function whose Walsh-Fourier series diverges everywhere.

The divergence of Fourier series on sets of measure zero has a long history. It is important to ask the question: for any set of measure zero, taken in advance, to construct a function whose Fourier series is divergent on the set. In fact the investigation of this problem began before Carleson's theorem. First, Stechkin [35] proved in 1951 that for any set $E \subset T$ of measure zero there exists a function $f \in L_2$ whose Fourier series diverges on E. Then, in 1963 Taukov [37] showed that f can be taken from L_p for any $1 \leq p < \infty$. In 1965 Kahane and Katznelson [16] proved the existence of a continuous complex valued function whose Fourier series diverges on a given set of measure zero. Essentially developing Kahane-Katznelson's approach Buzdalin [8] proved that for any set of measure zero there exists a continuous real valued function whose Fourier series diverges on that set. The same question has also been investigated for other classical orthonormal systems. Kheladze [19] (see also [20]) constructed a function from $L_p(1 < p < \infty)$ whose Walsh-Fourier series diverges on a given set of measure zero. Then Bugadze [6] proved that for the Walsh system the function in question can be taken from

L_∞. Goginava [13] proved that for any set of measure zero there exists a bounded function whose Walsh-Fejer means diverges on that set. Divergence of Fourier series in systems of characters of compact Abelian groups on a set of measure zero was studied by Oniani [26] and Bitsadze [5].

In the one-dimensional case the weak type inequality for the maximal operator of maximal operator of Fejér means can be found in Zygmund [39] for trigonometric series and in Schipp [31] for Walsh series. It follows (see [25, 29]) that the Fejér means $\sigma_n f$ of the Walsh-Fourier series of $f \in L_1$ converge to f almost everywhere.

Bugadze [7] proved that for any set of measure zero there exists a function in $f \in L_1(G)$ whose Féjer means of the Walsh-Fourier series diverges on the set. Karagulyan [17] (see also [18]) considered sequences of linear operators $\{U_n f(x)\}$ with some localization property, which is fulfilled for all summability methods with respect to orthonormal systems and proved that for any set E of measure zero there exists a set A for which $\{U_n I_A(x)\}$ diverges at each point $x \in E$, where I_A is a characteristic function of a subset A. This result is a generalization of analogous theorems known for the Fourier sum operators with respect to different orthogonal systems.

In this paper we construct an explicit function in $L_p(G)(1 \leq p < \infty)$ whose Féjer means with respect to the Walsh system diverges on any set of measure zero. In fact, the theorem we prove follows from the general result of [17], but we provide an alternative approach and the constructed function in our proof has a simple and explicit representation.

In Sect. 3 we present and prove our main result. Some corollaries and final remarks are given in Sect. 4. In order not to disturb our discussion in Sects. 1 and 3 we reserve Sect. 2 for some preliminaries.

2 Preliminaries

Denote by $\mathbb{N}_+$ the set of the positive integers, $\mathbb{N} := \mathbb{N}_+ \cup \{0\}$. Denote by $Z_2 := \{0, 1\}$ the additive group of integers modulo 2.

Define the group G as the complete direct product of the groups Z_2 with the product of the discrete topologies of Z_2's. The elements of G are represented by sequences $x := (x_0, x_1, \cdots, x_j, \cdots), (x_j \in Z_2)$.

The direct product μ of the measures $\mu_k(\{j\}) := 1/2 \ (j \in Z_2)$ is the Haar measure on G with $\mu(G) = 1$.

It is easy to give a base for the neighborhood of G :

$$I_0(x) := G, \quad I_n(x) := \{y \in G \mid y_0 = x_0, \cdots, y_{n-1} = x_{n-1}\},$$

where $x \in G$, $n \in \mathbb{N}$. Denote $I_n := I_n(0)$ for $n \in \mathbb{N}_+$, and $\overline{I_n} := G \setminus I_n$.

It is easy to show that every $n \in \mathbb{N}$ can be uniquely expressed as $n = \sum_{j=0}^{\infty} n_j 2^j$, where $n_j \in Z_2$ $(j \in \mathbb{N}_+)$ and only a finite numbers of n_j's differ from zero. For two natural numbers $n = \sum_{j=1}^{\infty} 2^j$ and $k = \sum_{j=1}^{\infty} k_j 2^j$, we define the operation $\oplus$ by $n \oplus k := \sum_{i=0}^{\infty} \left((n_j + k_j) \pmod{m_i}\right) 2^j, \quad n_j, k_j \in Z_2$.

It is known that (see [29])

$$w_n (x + y) = w_n (x)\; w_n (y), \tag{2.1}$$
$$w_{n \oplus k} (x) = w_k (x)\, w_n (x), \quad (k, n \in \mathbb{N}, \quad x, y \in G).$$

We define the complex-valued function $r_k (x) : G_m \to \mathbb{C}$, called the generalized Rademacher functions, by

$$r_k (x) := (-1)^{x_k}, \quad x \in G, \quad k \in \mathbb{N}.$$

Now, define the Walsh system $w := (w_n : n \in \mathbb{N})$ on G as:

$$w_n(x) := \prod_{k=0}^{\infty} r_k^{n_k} (x), \quad (n \in \mathbb{N}).$$

The Walsh system is orthonormal and complete in $L_2 (G)$ (see e.g. the books [1] and [33]).

If $f \in L_1 (G_m)$, we define the Fourier coefficients, partial sums of the Fourier series, Dirichlet kernels with respect to the Walsh system by

$$\widehat{f} (n) := \int_G f w_n d\mu, \quad (n \in \mathbb{N})$$

$$S_n f := \sum_{k=0}^{n-1} \widehat{f} (k)\, w_k \text{ and } D_n := \sum_{k=0}^{n-1} w_k, \quad (n \in \mathbb{N}_+)$$

respectively. Recall that (see e.g. [1, 29])

$$D_{2^n} (x) = \begin{cases} 2^n, \text{ if } x \in I_n, \\ 0, \text{ if } x \notin I_n. \end{cases} \tag{2.2}$$

For any $n \in \mathbb{N}_+$, the n-th Féjer means and kernels with respect to Walsh and trigonometric systems are respectively defined by

$$\sigma_n f := \frac{1}{n} \sum_{k=0}^{n-1} S_k f = \sum_{k=0}^{n-1} \left(1 - \frac{k}{n}\right) \widehat{f} (k)\, w_k, \quad K_n := \frac{1}{n} \sum_{k=0}^{n-1} D_k = \sum_{k=0}^{n-1} \left(1 - \frac{k}{n}\right) w_k.$$

We also define the maximal operators S^* and σ^* of the partial sums and Féjer means with respect to Walsh and trigonometric systems, respectively by

$$S^* f := \sup_{n\in\mathbb{N}} |S_n f| \quad \text{and} \quad \sigma^* f := \sup_{n\in\mathbb{N}} |\sigma_n f| .$$

3 Divergence of Fejér Means of Walsh-Fourier Series on the Sets of Measure Zero

Our main theorem reads:

Theorem 3.1 *Let $1 \leq p < \infty$ If $E \subseteq G$ is a set of measure zero, then there exists a function $f \in L_p(G)$ such that Féjer means with respect to the Walsh system diverge on this set.*

Proof We begin with a general remark. If $A \subseteq G$ is a finite union of dyadic intervals $I_1, I_2, \ldots, I_n$ for some $n \in \mathbb{N}_+$ and if N is any non-negative integer, then there exists a Walsh polynomial P such that, for some $i \geq N$, $P = \sum_{k=2^N}^{2^i-1} c_k w_k$, which satisfies

$$|P(x)| = 1, \; (x \in A) \text{ and } \int_G |P|^p \, d\mu = \mu(A).$$

Indeed, if $s := \max\{2^N, 1/\mu(I_j) : 1 \leq j \leq n\}$, then, in view of (2.2), we find that $P := \chi_A w_s$ is such a polynomial. In particular, for the character function $\chi_{I_n(y)}$ of any set $I_n(y)$ it can be written by explicit form $\chi_{I_n(y)} = D_{2^n}(x-y)/2^n$. According to (2.1) we can conclude that it is Walsh polynomial.

To prove the theorem, suppose $E \subseteq G$ satisfies that $\mu(E) = 0$. Cover E with intervals $(I_k, k \in \mathbb{N})$ such that $\sum_{k=0}^{\infty} \mu(I_k) < 1$ and each $x \in E$ belongs to infinitely many of the sets I_k. Set $n_0 := 0$ and choose integers $n_0 < n_1 < n_2 \ldots$ such that

$$\sum_{k=n_j}^{\infty} \mu(I_j) < 2^{-j} \;\; (j \in \mathbb{N}).$$

Apply the general remark above successively to the sets $A_j := \bigcup_{k=n_j}^{n_{j+1}-1} I_k \; (j \in \mathbb{N})$ to generate integers $\alpha_0 := 0 < \alpha_1 < \alpha_2 < \ldots$ and Vilenkin polynomials $P_0, P_1, \ldots$ such that $sp(P_j) := \{n \in \mathbb{N} : \widehat{P}_j \neq 0\} \subset [2^{\alpha_j}, 2^{\alpha_{j+1}})$:

$$P_j = \sum_{k=2^{\alpha_j}}^{2^{\alpha_{j+1}}-1} c_k^j w_k,$$

$$\|P_j\|_p^p = \mu(A_j) \leq 2^{-j} \tag{3.1}$$

and

$$\left|P_j(x)\right| = 1 \quad x \in A_j, \quad \text{for} \quad j \in \mathbb{N}. \tag{3.2}$$

Let $g := \sum_{j=1}^{\infty} P_j$ and choose integers $\beta_0 < \beta_1 < \beta_2 < \ldots$, such that $\alpha_j < \beta_j$, $j \in \mathbb{N}$ and

$$2^{3\beta_j} < \left(2^{\alpha_j} + 2^{\beta_j}\right)^3 < 2^{\alpha_j} + 2^{\beta_{j+1}} < 2^{\alpha_{j+1}} + 2^{\beta_{j+1}} < 2^{\beta_{j+1}+1}. \tag{3.3}$$

Setting

$$f := \sum_{j=1}^{\infty} w_{2^{\beta_{j+1}}} P_j$$

we observe by (3.1) that this series converges in $L_p(G)$ norm. Hence $f \in L_p(G)$ and this series is the Walsh-Fourier series of f. Moreover, since the spectra of the polynomials P_j are pairwise disjoint and

$$\widehat{f}(k) = \begin{cases} c_k^l, & \text{if } k = 2^{\beta_{l+1}} + 2^{\alpha_l}, \ldots, 2^{\beta_{l+1}} + 2^{\alpha_{l+1}} - 1, \ l \in \mathbb{N}_+, \\ 0, & \text{otherwise.} \end{cases}$$

for any $j \in \mathbb{N}_+$ we have that

$$\begin{aligned} &\sigma_{\left(2^{\alpha_{j+1}}+2^{\beta_{j+1}}\right)^3} f - \sigma_{2^{\alpha_j}+2^{\beta_{j+1}}} f \\ &= w_{2^{\beta_{j+1}}} P_j - \sum_{k=2^{\alpha_j}+2^{\beta_{j+1}}}^{2^{\alpha_{j+1}}+2^{\beta_{j+1}}-1} \frac{k}{\left(2^{\alpha_{j+1}} + 2^{\beta_{j+1}}\right)^3} c_k^j w_k \\ &+ \sum_{l=0}^{j-1} \sum_{k=2^{\alpha_l}+2^{\beta_{l+1}}}^{2^{\alpha_{l+1}}+2^{\beta_{l+1}}-1} \left(\frac{k}{2^{\alpha_j} + 2^{\beta_{j+1}}} - \frac{k}{\left(M_{\alpha_{j+1}} + 2^{\beta_{j+1}}\right)^3}\right) c_k^l w_k \\ &:= I - II + III. \end{aligned} \tag{3.4}$$

Since $\left|c_k^j\right| < C$, for any $j \in \mathbb{N}$, according to (3.3) for II we find that

$$\begin{aligned} |II| &\le C \sum_{k=0}^{M_{\alpha_{j+1}}+2^{\beta_{j+1}}-1} \frac{k}{\left(2^{\alpha_{j+1}} + 2^{\beta_{j+1}}\right)^3} \le \frac{C\left(2^{\alpha_{j+1}} + 2^{\beta_{j+1}}\right)^2}{\left(2^{\alpha_{j+1}} + 2^{\beta_{j+1}}\right)^3} \\ &\le \frac{C}{\left(2^{\alpha_{j+1}} + 2^{\beta_{j+1}}\right)} \to 0, \text{ as } j \to \infty. \end{aligned}$$

Analogously, if we apply again (3.3) we get that

$$|III| \leq \sum_{l=0}^{j-1} \sum_{k=2^{\alpha_l}+2^{\beta_l+1}}^{2^{\alpha_{l+1}}+2^{\beta_{l+1}}} \left(\frac{k}{2^{\alpha_j}+2^{\beta_{j+1}}} - \frac{k}{\left(2^{\alpha_{j+1}}+2^{\beta_{j+1}}\right)^3} \right) |c_k^l|$$

$$\leq C \sum_{k=0}^{2^{\alpha_j}+2^{\beta_j}} \frac{k}{2^{\alpha_j}+2^{\beta_{j+1}}} \leq C \sum_{k=0}^{2^{\beta_j+1}} \frac{k}{2^{\beta_{j+1}}} \leq \frac{C\left(2^{\beta_j}\right)^2}{2^{\beta_{j+1}}} \to 0, \text{ as } j \to \infty.$$

Since every $x \in E$ belongs to infinitely many of the sets A_j it follows from (3.2) that $|I| = 1$, for $x \in I_{A_j}$, for infinitely many j. Hence, we obtain that the Walsh-Fourier series of f diverges at every point $x \in E$. The proof is complete.

4 Corollary and Final Remarks

Remark 4.1 In fact, Theorem 3.1 follows from the general result of Karagulyan [17], but we provide an alternative approach and the constructed function in our proof has a simple and explicit representation.

Corollary 4.2 (See [28] and [33]) *Let $1 \leq p < \infty$. If $E \subseteq G$ is a set of measure zero, then there exists a function $f \in L_p(G)$ such that partial sums with respect to Walsh system diverge on this set.*

Remark 4.3 Almost everywhere convergence of some Nörlund and T means can be found in [29] (see also [3, 25, 27, 38]). We note that this type of construction in Theorem 3.1 can be used to prove similar results for Nörlund and T means.

References

1. G.N. Agaev, N.Y. Vilenkin, G.M. Dzhafarly, A.I. Rubinshtein, *Multiplicative Systems of Functions and Harmonic Analysis on Zero-dimensional Groups* (Elm, Baku, 1981) (Russian)
2. D. Baramidze, K. Nagy, L.E. Persson, G. Tephnadze, A sharp boundedness result concerning maximal operators of Vilenkin-Fourier series on martingale Hardy spaces. Georgian Math. J. **26**(3), 351–360 (2019)
3. D. Baramidze, N. Nadirashvili, L.-E. Persson, G. Tephnadze, Some weak-type inequalities and almost everywhere convergence of Vilenkin-Nörlund means. J. Inequal. Appl. **2023**, paper no. 66, 17 pp. (2023)
4. P. Billard, Sur la convergence presque partout des séries de Fourier-Walsh des fonctions de l'espace $L^2[0, 1]$. Stud. Math. **28**, 363–388 (1966/1967)
5. K. Bitsadze, On divergence of Fourier series with respect to multiplicative systems on the sets of measure zero. Georgian Math. J. **16**(3), 435–448 (2009)
6. V.M. Bugadze, Divergence of Fourier-Walsh series of bounded functions on sets of measure zero. Mat. Sb. **185**(7), 119–127 (1994) (in Russian); English transl.: Russian Math. Sb. **185**, 365–372 (1994)

7. V.M. Bugadze, On divergence of Fourier-Walsh series of bounded functions on sets of measure zero, Russian Academy of Sciences. Sb. Math. **82** (2), 365–372 (1995)
8. V.V. Buzdalin, Unboundedly diverging trigonometric Fourier series of continuous functions. Mat. Zametki **7**(1), 7–18 (1970) (in Russian); English transl.: Math. Notes **7**(1), 5–12 (1970)
9. L. Carleson, On convergence and growth of partial sums of Fourier series. Acta Math. **116**, 135–157 (1966)
10. C. Demeter, A guide to Carleson's theorem. Rocky Mountain J. Math. **45**(1), 169–212 (2015)
11. Ch. Fefferman, Pointwise convergence of Fourier series. Ann. Math. **98**(3), 551–571 (1973)
12. D.H. Fremlin, *Measure Theory*, vol. 2, Broad Foundations. Corrected second printing of the 2001 original (Torres Fremlin, Colchester, 2003)
13. U. Goginava, On divergence of Walsh-Fejer means of bounded functions on sets of measure zero. Acta Math. Hungar. **121**(3), 359–369 (2008)
14. L. Grafakos, *Classical Fourier Analysis*, 3rd edn, vol. 249. Graduate Texts in Mathematics (Springer, New York, 2014)
15. R. Hunt, *On the convergence of Fourier series, Orthogonal Expansions and their Continuous Analogues (Proc. Conf., Edwardsville, Ill., 1967)* (Southern Illinois Univ. Press, Carbondale, 1968), pp. 235–255
16. J.-P. Kahane, Y. Katznelson, Sur les ensembles de divergence des séries trigonométriques. Stud. Math. **26**, 305–306 (1966)
17. G.A. Karagulyan, Divergence of general operators on sets of measure zero. Colloquium Math. **121**(1), 113–119 (2010)
18. G.A. Karagulyan, On a characterization of the sets of divergence points of sequences of operators with the localization property (Russian). Mat. Sb., **202**(1), 11–36 (2011)
19. Sh.V. Kheladze, On the everywhere divergence of Fourier-Walsh series. Soobshch. Akad. Nauk. Gruzin. **77**, 305–307 (1975)
20. Sh.V. Kheladze, On everywhere divergence of Fourier series with respect to bounded type Vilenkin systems. Trudy Tbiliss. Mat. Inst. Gruzin. SSR (in Russian) **58**, 225–242 (1978)
21. A.N. Kolmogorov, Une série de Fourier-Lebesgue divergente presque partout. Polska Akad. Nauk. Fund. Math. **4**, 324–328 (1923)
22. M.T. Lacey, Carleson's theorem: proof, complements, variations. Publ. Mat. **48**(2), 251–307 (2004)
23. M. Lacey, C. Thiele, A proof of boundedness of the Carleson operator. Math. Res. Lett. **7**(4), 361–370 (2000)
24. N.N. Luzin, Collected works. In: *Metric Theory of Function and Functions of a Complex Variable*, vol. I. (Russian) (Izdat. Acad. Nauk. SSSR, Moscow, 1953)
25. N. Nadirashvili, L.-E. Persson, G. Tephnadze, F. Weisz, Vilenkin-Lebesgue points and almost everywhere convergence for some classical summability methods. Mediterr. J. Math. **19**, 239 (2022)
26. G.G. Oniani, On the divergence sets of Fourier series in systems of characters of compact abelian groups (Russian); translated from Mat. Zametki 112(1), 95–105 (2022). Math. Notes **112**(1–2), 100–108 (2022)
27. L.-E. Persson, G. Tephnadze, P. Wall, On the maximal operators of Vilenkin-Nörlund means. J. Fourier Anal. Appl. **21**(1), 76–94 (2015)
28. L.-E. Persson, F. Schipp, G. Tephnadze, F. Weisz, An analogy of the Carleson-Hunt theorem with respect to Vilenkin systems. J. Fourier Anal. Appl. **28**(48), 1–29 (2022)
29. L.-E. Persson, G. Tephnadze, F. Weisz, *Martingale Hardy Spaces and Summability of One-dimensional Vilenkin-Fourier Series* (Birkhäuser/Springer, 2022)
30. F. Schipp, Über die divergenz der Walsh-Fourierreihen. Ann. Univ. Sci. Budapest Eötvös Sect. Math. **12**, 49–62 (1969)
31. F. Schipp, Certain rearrangements of series in the Walsh series. Mat. Zametki **18**, 193–201 (1975)
32. F. Schipp, Pointwise convergence of expansions with respect to certain product systems. Anal. Math. **2**(1), 65–76 (1976)

33. F. Schipp, W.R. Wade, P. Simon, J. Pál, *Walsh series. An Introduction to Dyadic Harmonic Analysis* (Adam Hilger, Bristol, 1990)
34. P. Sjölin, An inequality of Paley and convergence a.e. of Walsh-Fourier series. Ark. Mat. **7**, 551–570 (1969)
35. S.B. Stechkin, On the convergence and divergence of trigonometric series. Uspekhi Mat. Nauk (in Russian) **6**(2), 148–149 (1951)
36. E.M. Stein, On limits of sequences of operators. Ann. Math. **74**(2), 140–170 (1961)
37. L.V. Taukov On the divergence of Fourier series with respect to a rearranged trigonometric system (in Russian). Uspekhi Mat. Nauk **18**(5), 191–198 (1963)
38. G. Tephnadze, Martingale Hardy spaces and summability of the one dimensional Vilenkin-Fourier series, Ph.D. thesis, Department of Engineering Sciences and Mathematics, Luleå University of Technology, ISSN 1402–1544, 2015
39. A. Zygmund, *Trigonometric Series*, vols. I, II (Cambridge University Press, 1959)

Martingale Hardy Spaces and Some Maximal Operators Associated with Walsh-Fejér Means

Davit Baramidze

Abstract In this paper we state and introduce some new restricted and weighted maximal operators of Fejér means with respect to Walsh-Fourier system for which (H_p, L_p) and $(H_p, weak - L_p)$ types inequalities hold for $0 < p \leq 1/2$. The sharpness of these results are also investigated. As a consequence we obtain both some new and well-know results. We note that these results are essential parts of the thesis (Baramidze, Martingale Hardy spaces and restricted and weighted maximal operators of Fejer means, PhD thesis, UiT The Arctic University of Norway, in preparation).

Keywords Walsh system · Fejér means · Martingale Hardy space · Maximal operators

2010 Mathematics Subject Classification 26015, 42C10, 42B30

1 Introduction

In order to be able to formulate some well-known and new results we need some preliminaries from the standard literature (see e.g. [32, 34]).

Let $\mathbb{N}_+$ denote the set of positive integers, $\mathbb{N} := \mathbb{N}_+ \cup \{0\}$. Denote by Z_2 the discrete cyclic group of order 2, that is, $Z_2 := \{0, 1\}$, where the group operation is

The research was supported by Shota Rustaveli National Science Foundation grant no. PHDF-21-1702.

D. Baramidze (✉)
School of Science and Technology, The University of Georgia, Tbilisi, Georgia

Department of Computer Science and Computational Engineering, UiT - The Arctic University of Norway, Narvik, Norway
e-mail: davit.baramidze@ug.edu.ge

R. Duduchava et al. (eds.), *Tbilisi Analysis and PDE Seminar*,
Research Perspectives Ghent Analysis and PDE Center 7,
https://doi.org/10.1007/978-3-031-62894-8_4

the modulo 2 addition and every subset is open. The Haar measure on Z_2 is given so that the measure of a singleton is $1/2$. Define the group G as the complete direct product of the group Z_2, with the product of the discrete topologies of Z_2.

We investigate the orthogonal Walsh system $\{w_n : n \in \mathbb{N}\}$ on the group G which are also group characters. If $f \in L_1$, then we can define Fourier coefficients, partial sums of Walsh-Fourier series and Fejér means as

$$\widehat{f}(n) := \int_G f w_n d\mu, \ (n \in \mathbb{N}), \ S_n f := \sum_{k=0}^{n-1} \widehat{f}(k) w_k,$$

$$\sigma_n f := \frac{1}{n} \sum_{k=1}^{n} S_k f, \quad (n \in \mathbb{N}_+),$$

The σ-algebra generated by the intervals $\{I_n(x) : x \in G\}$, where

$$I_0(x) := G, \ \ I_n(x) := \{y \in G : y_0 = x_0, \dots, y_{n-1} = x_{n-1}\} \ (n \in \mathbb{N}),$$

will be denoted by $\zeta_n \ (n \in \mathbb{N})$. Denote by $F = (F_n, n \in \mathbb{N})$ a martingale with respect to $\zeta_n \ (n \in \mathbb{N})$ (see e.g. [32, 46]). The maximal function F^* of a martingale F is defined by $F^* := \sup_{n\in\mathbb{N}} |F_n|$. For $0 < p < \infty$ the Hardy martingale spaces $H_p(G)$ consists of all martingales for which $\|F\|_{H_p} := \|F^*\|_p < \infty$.

It is easy to check that for every martingale $F = (F_n, n \in \mathbb{N})$ and every $k \in \mathbb{N}$ the limit $\widehat{F}(k) := \lim_{n\to\infty} \int_G F_n(x) w_k(x) d\mu(x)$ exists and it is called the k-th Walsh-Fourier coefficients of F.

A bounded measurable function a is a p-atom, if there exists an interval I, such that $\operatorname{supp}(a) \subset I$, $\int_I a d\mu = 0$ and $\|a\|_\infty \le \mu(I)^{-1/p}$. The atomic decomposition for the martingale Hardy spaces H_p, where $0 < p \le 1$, was proved by Weisz [48].

The weak (1, 1)-type inequality for the maximal operator σ^* of Fejér means σ_n with respect to the Walsh system, defined by $\sigma^* f := \sup_{n\in\mathbb{N}} |\sigma_n f|$, was derived in Schipp [33] and Pál, Simon [27] (see also [9, 26, 29]). Fujii [19] and Simon [35] verified that σ^* is bounded from $H_1(G)$ to $L_1(G)$. Weisz [47] generalized this result and proved boundedness of σ^* from the martingale space H_p to the Lebesgue space L_p for $p > 1/2$. Simon [36] gave a counterexample, which showed that boundedness does not hold for $0 < p < 1/2$. A counterexample for $p = 1/2$ was given by Goginava [21]. Moreover, in [22] (see also [37, 41–43, 45]) he proved that there exists a martingale $F \in H_p(G)$ $(0 < p \le 1/2)$, such that $\sup_{n\in\mathbb{N}} \|\sigma_n F\|_p = \infty$. Weisz [49] proved that the maximal operator σ^* of the Fejér means is bounded from the Hardy space $H_{1/2}$ to the space $weak - L_{1/2}$. Weisz [48] (see also [46]) also proved that for any $F \in H_p(G)$ $(p > 0)$, the maximal operator $\sup_{n\in\mathbb{N}} |\sigma_{2^n} F|$ is bounded from the Hardy space $H_p(G)$ to the Lebesgue space $L_p(G)$, but it is not bounded from the Hardy space $H_p(G)$ to the space $H_p(G)$ (see also [30, 32]).

In the study of the convergence of subsequences of Fejér means and their restricted maximal operators on the martingale Hardy spaces $H_p(G)$ for $0 < p \leq 1/2$, the central role is played by the fact that any natural number $n \in \mathbb{N}$ can be uniquely expressed as $n = \sum_{k=0}^{\infty} n_j 2^j, \quad n_j \in Z_2 \ (j \in \mathbb{N})$, where only a finite numbers of n_j differ from zero and their important characters $[n]$, $|n|$, $\rho(n)$ and $V(n)$ are defined by

$$[n] := \min\{j \in \mathbb{N}, n_j \neq 0\}, \quad |n| := \max\{j \in \mathbb{N}, n_j \neq 0\}, \quad \rho(n) = |n| - [n],$$

$$V(n) := n_0 + \sum_{k=1}^{\infty} |n_k - n_{k-1}|, \quad \text{for all} \quad n \in \mathbb{N}.$$

Moreover, for any natural numbers $\{n_{s_j}\}$, $j = 1, 2, \ldots, r$, satisfying $2^s \leq n_{s_1} \leq n_{s_2} \leq \ldots \leq n_{s_r} < 2^{s+1}$, $s \in \mathbb{N}$, we define numbers

$$s_- := \min\{[n_{s_j}]\}, \ s_+ := \max\{|n_{s_j}|\} = s \text{ and } \rho_s\left(n_{s_j}\right) := s_+ - s_-.$$

Furthermore, if $2^s \leq n_{s_j} \leq 2^{s+1}$, $j = 1, \ldots r$, $s \in \mathbb{N}$, which can be written as

$$n_{s_j} = \sum_{i=1}^{r_{s_j}} \sum_{k=l_i^{s_j}}^{t_i^{s_j}} 2^k, 0 \leq l_1^{s_j} \leq t_1^{s_j} \leq l_2^{s_j} - 2 < l_2^{s_j} \leq t_2^{s_j} \leq \ldots \leq l_{r_j}^{s_j} - 2 < l_{r_{s_j}}^{s_j} \leq t_{r_{s_j}}^{s_j},$$

we define

$$A_s := \left\{l_1^s, l_2^s, \ldots, l_{r_s^1}^s\right\} \bigcup \left\{t_1^s, t_2^s, \ldots, t_{r_s^2}^s\right\} = \left\{u_1^s, u_2^s, \ldots, u_{r_s^3}^s\right\},$$

where $u_1^s < u_2^s < \ldots < u_{r_s^3}^s$ and $t_{r_{s_j}}^{s_j} = s \in A_s$, for $j = 1, 2, \ldots, r$.

We also denote the cardinality of the set A_s by $|A_s|$, that is $card(A_s) := |A_s|$. By this definition we can conclude that $|A_s| = r_s^3 \leq r_s^1 + r_s^2$.

It is evident that $\sup_{s \in \mathbb{N}} |A_s| < \infty$ if and only if the sets $\{n_{s_1}, n_{s_2}, \ldots, n_{s_r}\}$ are uniformly finite for all $s \in \mathbb{N}_+$ and each n_{s_j} has bounded variation $V(n_{s_j})$, i.e. that

$$V(n_{s_j}) < c < \infty, \quad \text{for each} \quad j = 1, 2, \ldots, r.$$

Persson and Tephnadze [28] generalized this result and proved that if $0 < p \leq 1/2$ and $\{n_k : k \geq 0\}$ is a sequence of positive numbers, such that

$$\sup_{k \in \mathbb{N}} \rho(n_k) < c < \infty, \tag{1.1}$$

then the maximal operator $\widetilde{\sigma}^{*,\nabla}$, defined by

$$\widetilde{\sigma}^{*,\nabla} F := \sup_{k \in \mathbb{N}} \left|\sigma_{n_k} F\right|, \tag{1.2}$$

is bounded from the Hardy space $H_p(G)$ to the Lebesgue space $L_p(G)$. Moreover, if $0 < p < 1/2$ and $\{n_k : k \geq 0\}$ is a sequence of positive numbers, such that $\sup_{k\in\mathbb{N}} \rho(n_k) = \infty$, then there exists a martingale $F \in H_p(G)$ such that

$$\sup_{k\in\mathbb{N}} \left\|\sigma_{n_k} F\right\|_p = \infty.$$

In [44] it was proved that if $F \in H_{1/2}(G)$, then there exists an absolute constant c, such that

$$\|\sigma_n F\|_{H_{1/2}} \leq cV^2(n) \|F\|_{H_{1/2}} .$$

Moreover, if $\{n_k \in \mathbb{N} : k \geq 0\}$ is a sequence, such that $\sup_{k\in\mathbb{N}} V(n_k) = \infty$ and $\{\varphi_n\}$ is any nondecreasing, nonnegative function, satisfying the conditions $\Phi_n \uparrow \infty$ and $\overline{\lim_{k\to\infty}} \left(V^2(n_k)\Phi_{n_k}\right) = \infty$, then there exists a martingale $F \in H_{1/2}(G)$, such that

$$\sup_{k\in\mathbb{N}} \left\|\frac{\sigma_{n_k} F}{\Phi_{n_k}}\right\|_{1/2} = \infty.$$

It follows that if $f \in H_{1/2}(G)$ and $\{n_k : k \geq 0\}$ is any sequence of positive numbers, then $\sigma_{n_k} f$ are bounded from the Hardy space $H_{1/2}(G)$ to the space $H_{1/2}(G)$ if and only if, for some c, the following condition $\sup_{k\in\mathbb{N}} V(n_k) < c < \infty$ holds.

In [21] Goginava (see also [38]) proved that the weighted maximal operator $\overset{\sim}{\sigma}^*$, defined by

$$\overset{\sim}{\sigma}^* f := \sup_{n\in\mathbb{N}} \frac{|\sigma_n f|}{\log^2(n+1)}, \tag{1.3}$$

is bounded from the Hardy space $H_{1/2}(G)$ to the Lebesgue space $L_{1/2}(G)$. Moreover, it was also proved that the rate of increase of the denominator $\{\log^2(n+1)\}$ can not be improved. Similar problems for $0 < p < 1/2$ were investigated in [39, 40]. In particular, it was proved that the weighted maximal operator $\overset{\sim}{\sigma}^{*,p}$, defined by

$$\overset{\sim}{\sigma}^{*,p} f := \sup_{n\in\mathbb{N}} \frac{|\sigma_n f|}{(n+1)^{1/p-2}}, \quad (0 < p < 1/2) \tag{1.4}$$

is bounded from the Hardy space $H_p(G)$ to the Lebesgue space $L_p(G)$.

Similar problems for partial sums and some general summability methods can be found in [2], [5, 6], [7, 10–12], [15–18], [20], [23–25], and [31].

In this paper we introduce some new restricted and weighted maximal operators of Fejér means with respect to Walsh-Fourier system for which (H_p, L_p) and $(H_p, weak - L_p)$ types inequalities hold for $0 < p \leq 1/2$. The sharpness of these results are also investigated. As a consequence we obtain both some new and well-know results. We note that these results are essential parts of the thesis [3] where also the proofs of all statements and complementary information can be found.

2 The Main Result and Its Consequences

First we investigate case $p = 1/2$. This result was proved in [8].

Theorem 2.1

(a) *Let* $f \in H_{1/2}(G)$ *and* $\{n_k : k \geq 0\}$ *be a sequence of positive numbers and let* $\{n_{s_i} : 1 \leq i \leq r\} \subset \{n_k : k \geq 0\}$ *be numbers such that* $2^s \leq n_{s_1} \leq n_{s_2} \leq \ldots \leq n_{s_r} \leq 2^{s+1}$, $s \in \mathbb{N}$. *If the sets* A_s, *defined by* (1.1), *are uniformly finite for all* $s \in \mathbb{N}$, *that is the cardinality of the sets* A_s *are uniformly finite, that is* $\sup_{s\in\mathbb{N}}|A_s| < c < \infty$, *then the restricted maximal operator* $\widetilde{\sigma}^{*,\nabla}$, *defined by*

$$\widetilde{\sigma}^{*,\nabla} F = \sup_{k\in\mathbb{N}} \left|\sigma_{n_k} F\right|, \tag{2.1}$$

is bounded from the Hardy space $H_{1/2}$ *to the Lebesgue space* $L_{1/2}$.

(b) *(sharpness) Let* $\sup_{s\in\mathbb{N}}|A_s| = \infty$. *Then there exists a martingale* $f \in H_{1/2}(G)$, *such that the maximal operator, defined by* (2.1), *is not bounded from the Hardy space* $H_{1/2}$ *to the Lebesgue space* $L_{1/2}$.

In particular, Theorem 2.1 implies the following optimal characterization:

Corollary 2.2 *Let* $F \in H_{1/2}(G)$ *and* $\{n_k : k \geq 0\}$ *be a sequence of positive numbers. Then the restricted maximal operator* $\widetilde{\sigma}^{*,\nabla}$, *defined by* (2.1), *is bounded from the Hardy space* $H_{1/2}$ *to the Lebesgue space* $L_{1/2}$ *if and only if any sequence of positive numbers* $\{n_k : k \geq 0\}$ *which satisfies* $n_k \in [2^s, 2^{s+1})$, *is uniformly finite for each* $s \in \mathbb{N}_+$ *and each* $\{n_k : k \geq 0\}$ *has bounded variation, i.e.* $\sup_{k\in\mathbb{N}} V(n_k) < c < \infty$.

Our next theorem was proved in [1].

Theorem 2.3

(a) *Let* $f \in H_{1/2}$ *and* $\{n_k : k \geq 0\}$ *be a sequence of positive numbers. Then the weighted maximal operator* $\widetilde{\sigma}^{*,\nabla}$, *defined by*

$$\widetilde{\sigma}^{*,\nabla} F := \sup_{k\in\mathbb{N}} \frac{\left|\sigma_{n_k} F\right|}{\left|A_{|n_k|}\right|^2},$$

is bounded from the Hardy space $H_{1/2}$ *to the Lebesgue space* $L_{1/2}$.

(b) *(Sharpness) Let* $\sup_{k\in\mathbb{N}}|A_{|n_k|}| = \infty$ *and* $\{\varphi_n\}$ *be a nondecreasing sequence satisfying the condition*

$$\overline{\lim_{k\to\infty}} \frac{|A_{|n_k|}|^2}{\varphi_{|n_k|}} = \infty.$$

Then there exists a martingale $f \in H_{1/2}$, *such that the maximal operator, defined by*

$$\sup_{k\in\mathbb{N}} \frac{|\sigma_{n_k} F|}{\varphi_{|n_k|}}$$

is not bounded from the Hardy space $H_{1/2}$ *to the Lebesgue space* $L_{1/2}$.

From Theorem 2.3 we get the following result of Goginava [21] (see also [38]):

Corollary 2.4

(a) *The maximal operator* $\overset{\sim}{\sigma}^*$, *defined by* (1.3), *is bounded from the Hardy space* $H_{1/2}(G)$ *to the space* $L_{1/2}(G)$.
(b) *The rate of increase of the denominator* $\{\log^2(n+1)\}$ *can not be improved.*

In order to be able to compare with some other results in the literature we also state the following:

Corollary 2.5 *Let* $f \in H_{1/2}(G)$. *Then the restricted maximal operators* $\widetilde{\sigma}_i^{*,\nabla}$, $i = 1, 2, 3, 4, 5$, *defined by*

$$\widetilde{\sigma}_1^{*,\nabla} F := \sup_{k\in\mathbb{N}} \left|\sigma_{2^k} F\right|, \quad \widetilde{\sigma}_2^{*,\nabla} F := \sup_{k\in\mathbb{N}} \left|\sigma_{2^k+2^{k-1}} F\right|, \tag{2.2}$$

$$\widetilde{\sigma}_3^{*,\nabla} F := \sup_{k\in\mathbb{N}} \left|\sigma_{2^k+1} F\right|, \ \widetilde{\sigma}_4^{*,\nabla} F := \sup_{k\in\mathbb{N}} \left|\sigma_{2^{2k}+2^k} F\right|, \ \widetilde{\sigma}_5^{*,\nabla} F := \sup_{k\in\mathbb{N}} \left|\sigma_{2^{3k}+2^k} F\right|, \tag{2.3}$$

are bounded from the space $H_{1/2}(G)$ *to the Lebesgue space* $L_{1/2}(G)$.

Now we investigate $(H_p, weak - L_p)$ type inequalities for $0 < p < 1/2$ which are proved in [4].

Theorem 2.6

(a) *Let* $0 < p < 1/2$ *and* $f \in H_p(G)$. *Then the weighted maximal operator* $\overset{\sim}{\sigma}^{*,*,p}$ *defined by*

$$\overset{\sim}{\sigma}^{*,*,p} F := \sup_{n\in\mathbb{N}} \frac{|\sigma_n F|}{2^{\rho(n)(1/p-2)}},$$

is bounded from the Hardy space $H_p(G)$ *to the space weak* $- L_p(G)$.

(b) *Let* $0 < p < 1/2$ *and* $\varphi : \mathbb{N} \to [1, \infty)$ *be a nondecreasing function, satisfying the condition*

$$\overline{\lim_{n\to\infty}} \frac{2^{\rho(n)(1/p-2)}}{\varphi(n)} = \infty.$$

Then, there exist a sequence $\{f_{n_k}, k \in \mathbb{N}_+\}$ *of p-atoms and sequence* $\{q_{n_k}, k \in \mathbb{N}_+\}$ *of real numbers satisfying the condition* $|q_{n_k}| = n_k$, *such that*

$$\sup_{k\in\mathbb{N}} \frac{\left\| \frac{\sigma_{q_{n_k}} f_{n_k}}{\varphi(q_{n_k})} \right\|_{weak-L_p}}{\| f_{n_k} \|_{H_p}} = \infty.$$

In [4] it was also proved that the following theorem holds:

Theorem 2.7 *Let* $0 < p < 1/2$. *There exists a sequence* $\{f_k, k \in \mathbb{N}_+\}$ *of p-atoms, such that*

$$\sup_{k\in\mathbb{N}} \frac{\left\| \widetilde{\sigma}^{*,*,p} f_k \right\|_p}{\| f_k \|_{H_p}} = \infty.$$

In [13] (H_p, L_p) type inequalities were investigated for $0 < p < 1/2$ and the following sharp result was proved:

Theorem 2.8

(a) *Let* $0 < p < 1/2$, $f \in H_p(G)$ *and* $\varphi : \mathbb{N}_+ \to \mathbb{R}$ *be any nonnegative and nondecreasing function satisfying the condition*

$$\sum_{n=1}^{\infty} \frac{1}{\varphi^p(n)} < c < \infty. \tag{2.4}$$

Then the weighted maximal operator $\widetilde{\sigma}^{*,\nabla}$, *defined by*

$$\widetilde{\sigma}^{*,\nabla} F := \sup_{n\in\mathbb{N}} \frac{|\sigma_n F|}{2^{\rho(n)(1/p-2)}\varphi(\rho(n))},$$

is bounded from the Hardy space $H_p(G)$ *to the Lebesgue space* $L_p(G)$.

(b) *Let* $0 < p < 1/2$, $\{n_k : k \geq 0\}$ *be a sequence of positive numbers and* $\varphi : \mathbb{N}_+ \to \mathbb{R}$ *is any nonnegative and nondecreasing function satisfying the condition*

$$\sum_{n=1}^{\infty} \frac{1}{\varphi^p(n)} = \infty. \tag{2.5}$$

Then there exist p-atoms f_{n_k}, such that

$$\sup_{k\in\mathbb{N}} \frac{\left\| \sup_{n\in\mathbb{N}} \frac{|\sigma_n f_{n_k}|}{2^{\rho(n)(1/p-2)}\varphi(\rho(n))} \right\|_p}{\left\| f_{n_k} \right\|_{H_p}} = \infty.$$

If we take $\varphi(n) = n^{(1+\varepsilon)/p}$, for any $\varepsilon > 0$, we get that the condition (2.4) is fulfilled. On the other hand, if we take $\varphi(n) = n^{1/p}$, then the condition (2.5) holds. Hence, Theorem 2.8 imply the following sharp result:

Corollary 2.9

(a) *Let $0 < p < 1/2$ and $f \in H_p(G)$. Then the weighted maximal operator $\widetilde{\sigma}^{*,\nabla,\varepsilon}$, defined by*

$$\widetilde{\sigma}^{*,\nabla,\varepsilon} F := \sup_{n\in\mathbb{N}} \frac{|\sigma_n F|}{2^{\rho(n)(1/p-2)} (\rho(n))^{(1+\varepsilon)/p}}, \quad \varepsilon \geq 0,$$

is bounded from the Hardy space $H_p(G)$ to the Lebesgue space $L_p(G)$ for $\varepsilon > 0$ and is not bounded from the Hardy space $H_p(G)$ to the Lebesgue space $L_p(G)$ for $\varepsilon = 0$.

In [14] it was also proved that the following theorem holds:

Theorem 2.10

(a) *Let $0 < p < 1/2$, $f \in H_p(G)$ and $\{n_k : k \geq 0\}$ be a sequence of positive numbers and let $\left\{n_{s_i} : 1 \leq i \leq r\right\} \subset \{n_k : k \geq 0\}$ be numbers such that $2^s \leq n_{s_1} \leq n_{s_2} \leq \ldots \leq n_{s_r} \leq 2^{s+1}$, $s \in \mathbb{N}$. Then the weighted maximal operator $\widetilde{\sigma}^{*,\nabla}$, defined by*

$$\widetilde{\sigma}^{*,\nabla} F := \sup_{s\in\mathbb{N}} \sup_{2^s \leq n_{s_i} < 2^{s+1}} \frac{\left|\sigma_{n_{s_i}} F\right|}{2^{\rho_s(n_{s_i})(1/p-2)}},$$

is bounded from the Hardy space $H_p(G)$ to the Lebesgue space $L_p(G)$.

(b) *Let $0 < p < 1/2$, $f \in H_p(G)$, $\{n_k : k \geq 0\}$ be a sequence of positive numbers and let $\left\{n_{s_i} : 1 \leq i \leq r\right\} \subset \{n_k : k \geq 0\}$ be numbers such that $2^s \leq n_{s_1} \leq n_{s_2} \leq \ldots \leq n_{s_r} \leq 2^{s+1}$, $s \in \mathbb{N}$. Then, for any non-negative and non-decreasing function $\varphi : \mathbb{R}_+ \to \mathbb{R}$ satisfying the condition*

$$\sup_{s\in\mathbb{N}} \sup_{2^s \leq n_{s_i} < 2^{s+1}} \frac{2^{\rho_s(n_{s_i})(1/p-2)}}{\varphi(n_{s_i})} = \infty,$$

there exist p-atoms f_s, such that

$$\frac{\left\| \sup_{s\in\mathbb{N}} \sup_{2^s \le n_{s_i} < 2^{s+1}} \frac{\left|\sigma_{n_{s_i}} f_s\right|}{\varphi(n_{s_i})} \right\|_p}{\|f_s\|_{H_p}} \to \infty, \quad as\ s \to \infty.$$

From Theorem 2.8 follow immediately the following result of Tephnadze [39]:

Corollary 2.11

(a) *Let $0 < p < 1/2$ and $f \in H_p(G)$. Then the weighted maximal operator $\overset{\sim}{\sigma}^*$, defined by* (1.4), *is bounded from the martingale Hardy space $H_p(G)$ to the Lebesgue space $L_p(G)$.*
(b) *The rate of increase of the denominator $\{(n+1)^{1/p-2}\}$ can not be improved.*

From Theorem 2.10 we also obtain results of Persson and Tephnadze [28] (see also [9]):

Corollary 2.12 *If $0 < p < 1/2$, $f \in H_p(G)$ and $\{n_k : k \ge 0\}$ is any sequence of positive numbers, then the maximal operator defined by* (1.2) *is bounded from the Hardy space $H_p(G)$ to the Lebesgue space $L_p(G)$ if and only if the condition* (1.1) *is fulfilled.*

We also have the following new consequences of Theorem 2.8:

Corollary 2.13 *Let $p > 0$ and $f \in H_p(G)$. Then the restricted maximal operators $\widetilde{\sigma}_i^{*,\nabla}$, $i = 1, 2$, defined by* (2.2)*, are both bounded from the Hardy space $H_p(G)$ to the Lebesgue space $L_p(G)$.*

Corollary 2.14 *Let $0 < p < 1/2$ and $f \in H_p(G)$. Then the restricted maximal operators $\widetilde{\sigma}_i^{*,\nabla}$, $i = 3, 4, 5$, defined by* (2.3)*, are not bounded from the Hardy space $H_p(G)$ to the Lebesgue space $L_p(G)$.*

Corollary 2.15 *Let $0 < p < 1/2$ and $f \in H_p(G)$. Then the restricted maximal operator $\widetilde{\sigma}_i^{*,\nabla}$, $i = 6, 7, 8$, defined by*

$$\widetilde{\sigma}_6^{*,\nabla} F := \sup_{k\in\mathbb{N}} \frac{\left|\sigma_{2^k+1} F\right|}{2^{k(1/p-2)}}, \quad \widetilde{\sigma}_7^{*,\nabla} F := \sup_{k\in\mathbb{N}} \frac{\left|\sigma_{2^{2k}+2^k} F\right|}{2^{k(1/p-2)}}, \quad \widetilde{\sigma}_8^{*,\nabla} F := \sup_{k\in\mathbb{N}} \frac{\left|\sigma_{2^{3k}+2^k} F\right|}{2^{2k(1/p-2)}}, \tag{2.6}$$

are bounded from the Hardy space $H_p(G)$ to the Lebesgue space $L_p(G)$.

Remark 2.16 Let $0 < p < 1/2$ and $f \in H_p(G)$. Then from Corollary 2.5 we conclude that the restricted maximal operator $\widetilde{\sigma}_i^{*,\nabla}$, $i = 6, 7, 8$, defined by (2.6) are bounded from the space $H_{1/2}(G)$ to the Lebesgue space $L_{1/2}(G)$.

Acknowledgments I would like to thank professor Lars-Erik Persson for his very good suggestions, which have improved the paper. The research was supported by The University of Georgia grant "Mathematical Analysis and PDE on Lie Groups."

References

1. N. Areshidze, D. Baramidze, L.-E. Persson, G. Tephnadze, Some new weighted maximal operators of Fejér means of Walsh-Fourier series, Anal. Appl., to appear
2. D. Baramidze, Sharp $\left(H_p, L_p\right)$ and $\left(H_p, \text{weak} - L_p\right)$ type inequalities of weighted maximal operators of T means with respect to Vilenkin systems. Trans. A. Razmadze Math. Inst. **176**(2), 173–182 (2022)
3. D. Baramidze, Martingale Hardy spaces and restricted and weighted maximal operators of Fejer means, PhD thesis, UiT The Arctic University of Norway, in preparation
4. D. Baramidze, G. Tephnadze, Some new weak-$(H_p - L_p)$ type inequalities for weighted maximal operators of Fejér means of Walsh-Fourier series. Acta Math. Hungar. **171**(2), 267–283 (2023)
5. D. Baramidze, Z. Dvalashvili, G. Tutberidze, Convergence of Nörlund means with respect to Vilenkin systems of integrable functions. Mem. Differential Equations Math. Phys. **86**, 1–14 (2022)
6. D. Baramidze, N. Gogolashvili, N. Nadirashvili, Convergence of T means with respect to Vilenkin systems of integrable functions. Georgian Math. J. **29**, 481–491 (2022)
7. D. Baramidze, L.-E. Persson, H. Singh, G. Tephnadze, Some new results and inequalities for subsequences of Nörlund logarithmic means of Walsh-Fourier series. J. Inequal. Appl. **2022**, paper no. 30, 13 pp. (2022)
8. D. Baramidze, L. Baramidze, L.-E. Persson, G. Tephnadze, Some now restricted maximal operators of Fejér means of Walsh-Fourier series. Banach J. Math. Anal. **17**(4), paper no. 75, 20 pp. (2023)
9. D. Baramidze, N. Nadirashvili, L.-E. Persson, G. Tephnadze, Some weak-type inequalities and almost everywhere convergence of Vilenkin-Nörlund means. J. Inequal. Appl. **2023**, paper no. 66, 17 pp. (2023)
10. D. Baramidze, L.-E. Persson, K. Tangrand, G. Tephnadze, $(H_p - L_p)$ type inequalities for subsequences of Nörlund means of Walsh-Fourier series. J. Inequal. Appl. **2023**, paper no. 52, 15 pp. (2023)
11. D. Baramidze, L.-E. Persson, H. Singh, G. Tephnadze, Some new weak $(H_p - L_p)$ type inequality for weighted maximal operators of partial sums of Walsh-Fourier series. Mediterr. J. Math. **20**(5), paper no. 28, 13 pp. (2023)
12. D. Baramidze, L.-E. Persson, G. Tephnadze, Some new $(H_p - L_p)$ type inequalities for weighted maximal operators of partial sums of Walsh-Fourier series. Positivity **27**(3), paper no. 38, 14 pp. (2023)
13. D. Baramidze, I. Blahota, G. Tephnadze, R. Toledo, Martingale Hardy spaces and some new weighted maximal operator of Fejér means of Walsh-Fourier series. J. Geom. Anal. **34**(1), 17 pp. (2024)
14. D. Baramidze, A. Meidell, D. Lukkassen, L.-E. Persson, Some new $(H_p - L_p)$ type inequalities for weighted maximal operators of Fejér means of Walsh-Fourier series. Trans. A. Razmadze Math. Inst., to appear
15. I. Blahota, G. Tephnadze, On the (C, α)-means with respect to the Walsh system. Anal. Math. **40**, 161–174 (2014)
16. I. Blahota, G. Tephnadze, Strong convergence theorem for Vilenkin-Fejér means. Publ. Math. Debrecen **85**(1–2), 181–196 (2014)
17. I. Blahota, G. Tephnadze, R. Toledo, Strong convergence theorem of (C, α)-means with respect to the Walsh system. Tohoku Math. J. **67**(4), 573–584 (2015)

18. I. Blahota, K. Nagy, G. Tephnadze, Approximation by Marcinkiewicz Θ-means of double Walsh-Fourier series. Math. Inequal. Appl. **22**(3), 837–853 (2019)
19. N.J. Fujii, A maximal inequality for H_1 functions on the generalized Walsh-Paley group. Proc. Amer. Math. Soc. **77**, 111–116 (1979)
20. G. Gát, Investigations of certain operators with respect to the Vilenkin system. Acta Math. Hungar. **61**, 131–149 (1993)
21. U. Goginava, Maximal operators of Fejér-Walsh means. Acta Sci. Math. (Szeged) **74**(3–4), 615–624 (2008)
22. U. Goginava, The martingale Hardy type inequality for Marcinkiewicz-Fejér means of two-dimensional conjugate Walsh-Fourier series. Acta Math. Sin. **27**, 1949–1958 (2011)
23. N. Gogolashvili, K. Nagy, G. Tephnadze, Strong convergence theorem for Walsh-Kaczmarz-Fejér means. Mediterr. J. Math. **18**(2), paper no. 37, 17 (2021)
24. D. Lukkassen, L.E. Persson, G. Tephnadze, G. Tutberidze, Some inequalities related to strong convergence of Riesz logarithmic means of Vilenkin-Fourier series. J. Inequal. Appl. **2020**, paper no. 79, 17 pp. (2020)
25. N. Nadirashvili, L.-E. Persson, G. Tephnadze, F. Weisz, Vilenkin-Lebesgue points and almost everywhere convergence of Vilenkin-Fejér means and applications. Mediterr. J. Math. **19**(5), paper no. 239, 16 pp. (2022)
26. N. Nadirashvili, G. Tephnadze, G. Tutberidze, Almost everywhere and norm convergence of approximate identity and Fejér means of trigonometric and Vilenkin systems. Trans. A. Razmadze Math. Inst. **177**(3), 453–463 (2023)
27. J. Pál , P. Simon, On a generalization of the concept of derivate. Acta Math. Hung. **29**, 155–164 (1977)
28. L.E. Persson, G. Tephnadze, A sharp boundedness result concerning some maximal operators of Vilenkin-Fejér means. Mediterr. J. Math. **13**(4), 1841–1853 (2016)
29. L.E. Persson, G. Tephnadze, P. Wall, On the maximal operators of Vilenkin-Nörlund means. J. Fourier Anal. Appl. **21**(1), 76–94 (2015)
30. L.E. Persson, G. Tephnadze, G. Tutberidze, On the boundedness of subsequences of Vilenkin-Fejér means on the martingale Hardy spaces. Oper. Matrices **14**(1), 283–294 (2020)
31. L.-E. Persson, F. Schipp, G. Tephnadze, F. Weisz, An analogy of the Carleson-Hunt theorem with respect to Vilenkin systems. J. Fourier Anal. Appl. **28**(3), paper no. 48, 29 pp. (2022)
32. L.E. Persson, G. Tephnadze, F. Weisz, *Martingale Hardy Spaces and Summability of One-dimensional Vilenkin-Fourier Series* (Birkhäuser/Springer, 2022)
33. F. Schipp, Certain rearrangements of series in the Walsh series. Mat. Zametki **18**, 193–201 (1975)
34. F. Schipp, W. Wade, P. Simon, J. Pál, *Walsh Series, An Introduction to Dyadic Harmonic Analysis* (Akademiai Kiadó, Budapest-Adam Hilger, 1990)
35. P. Simon, Investigations with respect to the Vilenkin system. Ann. Univ. Sci. Budapest Eötvös Sect. Math. **28**, 87–101 (1985)
36. P. Simon, Cesáro summability with respect to two-parameter Walsh systems. Monatsh. Math. **131**(4), 321–334 (2000)
37. G. Tephnadze, Fejér means of Vilenkin-Fourier series. Stud. Sci. Math. Hungar. **49**, 79–90 (2012)
38. G. Tephnadze, On the maximal operator of Vilenkin-Fejér means. Turk. J. Math. **37**, 308–318 (2013)
39. G. Tephnadze, On the maximal operators of Vilenkin-Fejér means on Hardy spaces. Math. Inequal. Appl. **16**(2), 301–312 (2013)
40. G. Tephnadze, On the maximal operators of Walsh-Kaczmarz-Fejér means. Period. Math. Hung. **67**(1), 33–45 (2013)
41. G. Tephnadze, A note on the norm convergence by Vilenkin-Fejér means. Georgian Math. J. **21**(4), 511–517 (2014)
42. G. Tephnadze, Approximation by Walsh-Kaczmarz-Fejér means on the Hardy space. Acta Math. Sci. **34**(5), 1593–1602 (2014)

43. G. Tephnadze, Strong convergence theorems of Walsh-Fejér means. Acta Math. Hungar. **142**(1), 244–259 (2014)
44. G. Tephnadze, On the convergence of Fejér means of Walsh-Fourier series in the space H_p. J. Contemp. Math. Anal. **51**(2), 90–102 (2016)
45. G. Tephnadze, The one-dimensional martingale Hardy spaces and partial sums and Fejér means with respect to Walsh system. Mem. Differential Equations Math. Phys. **88**, 109–158 (2023)
46. F. Weisz, *Martingale Hardy Spaces and their Applications in Fourier Analysis* (Springer, Berlin, 1994)
47. F. Weisz, Cesáro summability of one- and two-dimensional Walsh-Fourier series. Anal. Math. **22**, 229–242 (1996)
48. F. Weisz, *Summability of Multi-dimensional Fourier Series and Hardy Space* (Kluwer Academic, Dordrecht, 2002)
49. F. Weisz, Weak type inequalities for the Walsh and bounded Ciesielski systems. Anal. Math. **30**(2), 147–160 (2004)

Generic Bessel Potential Spaces on Lie Groups

Duván Cardona, Roland Duduchava, Arne Hendrickx, and Michael Ruzhansky

Abstract The purpose of the present work is to define suitable (generic) Bessel Potential Spaces (GBPS) for convolution integro-differential equations (CIDE) on a Lie group G, which is homomorphic to the Euclidean space $\mathbb{R}^n$. Is proved that ellipticity of the symbol is necessary condition for the Fredholm property of all CIDE in GBPS, but not sufficient. Ellipticity is sufficient condition for unique solvability of CIDE for $p = 2$ and, for $p \neq 2$, only for CIDE with symbols from the subalgebra of $\mathbb{L}_p$-multipliers.

Keywords Lie group · Convolution on Lie group · Generic differential operators · Generic Bessel potential spaces

Mathematics Subject Classification 35J57, 45E10, 47B35

D. Cardona · A. Hendrickx
Department of Mathematics: Analysis, Logic and Discrete Mathematics, Ghent University, Ghent, Belgium
e-mail: duvan.cardonasanchez@ugent.be; arnhendr.Hendrickx@UGent.be

R. Duduchava (✉)
Institute of Mathematics, University of Georgia & A. Razmadze Mathematical Institute. I. Javakhishvili State Tbilisi, Tbilisi, Georgia
e-mail: r.duduchava@ug.edu.ge

M. Ruzhansky
Department of Mathematics: Analysis, Logic and Discrete Mathematics, Ghent University, Ghent, Belgium

School of Mathematical Sciences, Queen Mary University of London, London, United Kingdom
e-mail: michael.ruzhansky@ugent.be; m.ruzhansky@qmul.ac.uk

R. Duduchava et al. (eds.), *Tbilisi Analysis and PDE Seminar*, Research Perspectives Ghent Analysis and PDE Center 7,
https://doi.org/10.1007/978-3-031-62894-8_5

1 Convolutions on Lie Groups

We concentrate investigation on a Lie group $\{G, x \circ y\}$ with the group operation $x \circ y$, which is homomorphic to the Lie group $\{\mathbb{R}^n, x \circ y = x + y\}$ (actually in our examples we deal with infinitely smooth diffeomorphisms). Then on $\{G, x \circ y\}$ we have uniquely defined Haar measure $d_G\mu$, the Fourier transformation $\mathcal{F}_G$, the inverse Fourier transformation $\mathcal{F}_G^{-1}$ and generic differential operators $\mathfrak{D}_1, \dots, \mathfrak{D}_n$, generated by the vector fields from the corresponding Lie algebra. The dual group is then $\widehat{G} = \mathbb{R}^n$ and, using pull-back operators associated to the group homomorphism $x(t) : G \to \mathbb{R}^n$ and its inverse $t(x) : \mathbb{R}^n \to G$, we define the Schwartz spaces of fast decaying smooth functions $\mathbb{S}(G)$ and its dual-the space of Schwartz distributions $\mathbb{S}'(G)$.

More details about Lie groups can be retrieved, e.g., from the book [4].

For a (complex valued) symbol $a(\xi)$ from the space of Schwartz distributions on the dual space of characters $a \in \mathbb{S}'(\widehat{G}) = \mathbb{S}'(\mathbb{R}^n)$ we define the associated convolution operator as follows:

$$W_{a,G}^0 := \mathcal{F}_G^{-1} a \mathcal{F}_G : \mathbb{S}(G) \to \mathbb{S}'(G), \tag{1.1}$$

By $\mathfrak{M}_p(\widehat{G})$ we denote the algebra of functions such that the convolution operator

$$W_{a,G}^0 : \mathbb{L}_p(G, \boldsymbol{d}_G\mu) \to \mathbb{L}_p(G, \boldsymbol{d}_G\mu), \qquad 1 < p < \infty,$$

is bounded. Here $\mathbb{L}_p(G, \boldsymbol{d}_G\mu)$, $1 < p < \infty$ is the Lebesgue space with the measure $\boldsymbol{d}_G\mu$ of complex-valued functions. $\mathfrak{M}_p(\widehat{G})$ coincides, obviously, with the classical algebra of $\mathbb{L}_p$-multipliers on the Euclidean space $\mathfrak{M}_p(\mathbb{R}^n)$. The immediate consequence of the Plancherel's theorem on the isometrical isomorphism of the space $\mathcal{F}_G : \mathbb{L}_2(G) \to \mathbb{L}_2(G)$ is that $\mathcal{M}_2(\widehat{G}) = \mathbb{L}_\infty(\mathbb{R}^n)$, i.e., the convolution operator $W_{a,G}^0$ is bounded in $\mathbb{L}_2(G, \boldsymbol{d}_G\mu)$ if and only if $a(\xi)$ is an essentially bounded function on $\widehat{G} = \mathbb{R}^n$. It is well known, that $\mathfrak{M}_p(\mathbb{R}^n) \subset \mathfrak{M}_2(\mathbb{R}^n) = \mathbb{L}_\infty(\mathbb{R}^n)$.

By $\boldsymbol{PC}_p(\widehat{G}) = \boldsymbol{PC}_p(\mathbb{R}^n)$ we denote the closure of the algebra of piecewise-constant functions (constant on finite set of non-intersecting open polytopes (see [1]) in the norm of multipliers $\mathfrak{M}_p(\mathbb{R}^n)$. Due to the inequality (cf. [6]) $\|a|\mathfrak{M}_p(\widehat{G})\| = \|W_{G,a}^0|\mathbb{L}_p(G, d_G\mu)\| \geqslant \|W_{G,a}^0|\mathbb{L}_2(G, d_G\mu)\| = \sup_{\xi\in\widehat{G}} |a(\xi)|$ functions from $\boldsymbol{PC}_p(\mathbb{R}^n)$ have at most countable number of different radial limits at all points $x \in \mathbb{R}^n \bigcup\{\infty\}$.

For the class of matrix symbols and vector spaces we will use the same notations $\mathfrak{M}_p(\widehat{G})$, $\boldsymbol{PC}_p(\widehat{G})$ instead of $\mathfrak{M}_p^{m\times m}(\widehat{G})$, $\boldsymbol{PC}_p^{m\times m}(\widehat{G})$ and $\mathbb{L}_p(G, \boldsymbol{d}_G\mu)$, $\mathbb{GH}_p^s(G, \boldsymbol{d}_G\mu)$ (see later) instead of $\mathbb{L}_p^m(G, \boldsymbol{d}_G\mu)$, $(\mathbb{GH}_p^s)^m(G, \boldsymbol{d}_G\mu)$, respectively.

Theorem 1.1 *Let $1 < p < \infty$. A convolution equation*

$$\boldsymbol{W}^0_{G,a}\varphi(x) := c_0\varphi(x) + \int_G k(x \circ y^{-1})\varphi(y)\boldsymbol{d}_G\mu(y) = f(y), \tag{1.2}$$

$$\varphi, f \in \mathbb{L}_p(G, \boldsymbol{d}_G\mu), \quad a(\xi) = c_0 + (\mathcal{F}_G k)(\xi), \quad \xi \in \widehat{G}, \quad a \in \mathcal{M}_p(\widehat{G}),$$

where $k(x)$, $a(\xi)$ are $m \times m$ matrix-functions and $\varphi(x)$, $f(x)$ are m-vectors, is Fredholm only if the symbol is elliptic

$$\inf_{\xi \in \widehat{G}} |\det a(\xi)| > 0. \tag{1.3}$$

Moreover, if $a \in \boldsymbol{PC}_p(\widehat{G})$, the ellipticity (1.3) *is sufficient for equation* (1.2) *to have a unique solution $\varphi = \boldsymbol{W}^0_{G,a^{-1}} f \in \mathbb{L}_p(G, \boldsymbol{d}_G\mu)$ for arbitrary $f \in \mathbb{L}_p(G, \boldsymbol{d}_G\mu)$.*

Proof If $a \in \mathfrak{M}_p(\mathbb{R}^n)$ and $W^0_{G,a}$ is Fredholm in $\mathbb{L}_p(\mathbb{R}^n)$, due to the invariance with respect to shifts, it is invertible (see similar proofs in [2, 8]). Then the dual operator $(W^0_{G,a})^* = W^0_{G,\overline{a}} = \boldsymbol{V} W^0_{G,a}\boldsymbol{V}$, where $\boldsymbol{V}g(\xi) := \overline{g(\xi)}$ is the complex conjugation, is invertible in the dual space $\mathbb{L}_{p'}(\mathbb{R}^n)$, $p' := p/(p-1)$. Since $\boldsymbol{V}^2 = I$, the operator $W^0_{G,a}$ is invertible in the dual space $\mathbb{L}_{p'}(\mathbb{R}^n)$. Then, by interpolation, $W^0_{G,a}$ is invertible in $\mathbb{L}_2(\mathbb{R}^n)$ (note, that $2 \in [p, p']$). Due to isometrical isomorphism $\mathcal{F}_G : \mathbb{L}_2(G) \to \mathbb{L}_2(G)$ the operator $W^0_{G,a}$ is equivalent to the multiplication by the symbol $\mathcal{F}_G W^0_{G,a}\mathcal{F}_G^{-1} = aI$; thence, the multiplication operator by the symbol $aI : \mathbb{L}_2(\mathbb{R}^n) \to \mathbb{L}_2(\mathbb{R}^n)$ is invertible and the ellipticity of $a \in \mathbb{L}_\infty(\mathbb{R}^n)$ follows.

Let $\overset{\circ}{\mathbb{R}}{}^n$ be the compactification of $\mathbb{R}^n$ at infinity by the points of the unit sphere $S^{n-1}(0) := \{\omega \in \mathbb{R}^n : |\omega| = 1\}$ along the rays $\mathbb{R}_{\infty,\omega}$, emerging from 0 and crossing $\omega \in S^{n-1}(0)$.

For the proof of the second part we apply Gohberg-Krupnik's local principle (see [5, Vol. I, Ch. 5]): Choose localizing class $\boldsymbol{V}_{x,\zeta} := \{W^0_{G,v_{\zeta,x}}\}$ for $x \in \overset{\circ}{\mathbb{R}}{}^n$, $\zeta \in S^{n-1}(x)$, where $v_{x,\zeta}(\xi)$ is a characteristic function of the intersection of neighbourhoods of x and sector of the cone including $\zeta \in S^{n-1}(x)$ and with the vertex at x.

The multiplier $a \in \boldsymbol{PC}_p(\widehat{G})$ is approximated by piecewise-constant functions $a_\varepsilon \in \bigcup_{1<r<\infty}\mathfrak{M}_r(\widehat{G})$ and, due to the Hörmander's inequality (cf. [6]) $\|W^0_{G,a}|\mathbb{L}_p(G)\| \leqslant \sup_{\xi\in\mathbb{R}^n}|a(\xi)|^\theta \|W^0_{G,a}|\mathbb{L}_r(G)\|^{1-\theta}$, $0 < \theta < 1$, the local equivalence $W^0_{G,a} \overset{\boldsymbol{V}_{x,\zeta}}{\sim} W^0_{G,a(x,\zeta)} = a(x,\zeta)I$ is proved, where $a(x,\zeta)$ is the constant radial limit

$$a(x,\zeta) := \lim_{\xi\to x,\ \xi\in\mathbb{R}_{x,\zeta}} a(\xi), \qquad (x,\zeta) \in \overset{\circ}{\mathbb{R}}{}^n \times S^{n-1}(x).$$

Since the set of localizing classes $\{V_{x,\zeta}\}_{(x,\zeta)\in \overset{\circ}{\mathbb{R}^n}\times S^{n-1}}$ is covering, by the local principle $W^0_{G,a}$ is invertible if the local representatives $a(x,\zeta)I$ are invertible $\forall\,(x,\zeta)\in \overset{\circ}{\mathbb{R}^n}\times S^{n-1}$, i.e., if (1.3) holds. ■

Remark 1.1 The ellipticity (1.3) is not sufficient for the Fredholm property of (1.3) when $a\in\mathfrak{M}_p(\mathbb{R}^n)$, except $p=2$, due to the "Igari paradox": for any $p\neq 2$ there exists a continuous multiplier $a\in\mathfrak{M}_p(\mathbb{R}^n)$, which is elliptic $\inf_{\xi\in\mathbb{R}^n}|a(\xi)|\geqslant 1$ such that the inverse $a^{-1}(\xi):=1/a(\xi)$ is not a multiplier $a\notin\mathfrak{M}_p(\mathbb{R}^n)$ (cf. [7]).

2 Generic Bessel Potential Spaces

Since the Lie groups G we consider are homomorphic to $\mathbb{R}^m$, the dual groups $\widehat{G}$ coincide with the Euclidean space $\mathbb{R}^m$. Then the generic differential operators are

$$\mathfrak{D}_k := \boldsymbol{W}^0_{G,-i\xi_k} = \mathcal{F}_G^{-1}(-i\xi_k)\mathcal{F}_G, \qquad k=1,2,\dots,m. \tag{2.1}$$

Arises the following legal question: which spaces are suitable to consider solvability of the convolution integro-differential equation

$$\sum_{|\alpha|+|\beta|,|\gamma|\leqslant n}\left[c_{\alpha,\beta}\mathfrak{D}^\gamma\varphi(x)+\mathfrak{D}^\alpha\int_G k_{\alpha,\beta}(x\circ y^{-1})\mathfrak{D}^\beta\varphi(y)\boldsymbol{d}_G\mu(y)\right]=f(y) \tag{2.2}$$

with the symbol of polynomial growth $a(\xi) := \sum_{|\alpha|+|\beta|,|\gamma|\leqslant n}\big[c_{\alpha,\beta}(-i\xi)^\gamma+(-i\xi)^{\alpha+\beta}$ $\times\mathcal{F}_G k_{\alpha,\beta}(\xi)\big]$, $\xi\in\mathbb{R}^m$.

For this purpose we define the Generic Bessel Potential Spaces (GBPS) $\mathbb{GH}^s_p(G,\boldsymbol{d}_G\mu)$ on the Lie group G, endowed with the norm

$$\begin{aligned}\|\psi\mid\mathbb{GH}^s_p(G,\boldsymbol{d}_G\mu)\| &:= \|W^0_{G,\langle\xi\rangle^s}\psi\mid\mathbb{L}_p(G,\boldsymbol{d}_G\mu)\|\\ &= \|\mathcal{F}_G^{-1}\langle\xi\rangle^s\mathcal{F}_G\psi\mid\mathbb{L}_p(G,\boldsymbol{d}_G\mu)\| \quad \langle\xi\rangle^s:=(1+|\xi|^2)^s, \quad s\in\mathbb{R}.\end{aligned} \tag{2.3}$$

For an integer $s=m=1,2,\dots$ the space $\mathbb{GH}^m_p(G,\boldsymbol{d}_G\mu)$ is isomorphic to the Generic Sobolev Space (GSS) $\mathbb{GW}^m_p(G,\boldsymbol{d}_G\mu)$, consisting of functions with the finite norm

$$\|\varphi\mid\mathbb{GW}^m_p(G,\boldsymbol{d}_G\mu)\| := \left[\sum_{|\alpha|\leqslant m}\|\mathfrak{D}^\alpha\varphi\mid\mathbb{L}_p(G,\boldsymbol{d}_G\mu)\|^p\right]^{1/p}.$$

Theorem 2.1 *Let $1 < p < \infty$, $s \in \mathbb{R}$. An $m \times m$ system of convolution integro-differential equation* (2.2) *with the $m \times m$ matrix symbol $a \in \mathcal{M}_p(\widehat{G})$ is Fredholm in the setting $\varphi \in \mathbb{GH}^s_p(G, \boldsymbol{d}_G\mu)$, $f \in \mathbb{GH}^{s-m}_p(G, \boldsymbol{d}_G\mu)$ if only the symbol is elliptic*

$$\inf_{\xi\in\widehat{G}} |\det\left[\langle\xi\rangle^{-m}a(\xi)\right]| > 0. \tag{2.4}$$

Moreover, if $a \in \boldsymbol{CP}_p(\widehat{G})$, the ellipticity (2.4) *is sufficient for Eq.* (2.2) *to have a unique solution $\varphi = \mathbf{W}^0_{G,a^{-1}}f \in \mathbb{GH}^s_p(G, \boldsymbol{d}_G\mu)$ for arbitrary $f \in \mathbb{GH}^{s-m}_p(G, \boldsymbol{d}_G\mu)$.*

Remark 2.2 Theorems 1.1 and 2.1 are valid also for symbols (matrix symbols) from the Wiener algebra $W(\widehat{G}) := \left\{a(\xi) = c + (\mathcal{F}_G k)(\xi) \ : \ k \in \mathbb{L}_1(\widehat{G}, d_G\mu)\right\}$.

3 Examples

Example 3.1 The first example is the classical Euclidean space $G = \mathbb{R}^n$ with the group operation $x \circ y = x + y$, the dual group $\widehat{\mathbb{R}^n} = \mathbb{R}^n$, the Haar-Lebesgue measure $\boldsymbol{d}x$, the classical Fourier transformation and its inverse

$$(\mathcal{F}\varphi)(\xi) = \int_{\mathbb{R}^n} e^{i\xi x}\varphi(x)\boldsymbol{d}x, \quad \xi \in \mathbb{R}^n.$$
$$(\mathcal{F}^{-1}\psi)(x) = \frac{1}{(2\pi)^n}\int_{\mathbb{R}^n} e^{-i\xi x}\psi(\xi)d\xi, \quad x \in \mathbb{R}^n.$$

Convolution equation is classical

$$\mathbf{W}^0_a\varphi(x) := c_0\varphi(x) + \int_{\mathbb{R}^n} k(x-y)\varphi(y)\boldsymbol{d}y = f(y).$$

and partial differential operators are generic

$$\mathfrak{D}_k = \partial_k = \mathbf{W}^0_{-i\xi_k}, \qquad k = 1, 2, \ldots, n.$$

Example 3.2 The direct product of positive half axes $\mathbb{R}^n_+ = (0, \infty)^n$ is a Lie group with the group operation $x \circ y = (x_1y_1, \ldots, x_ny_n)$, with the dual $\widehat{\mathbb{R}^n_+} = \mathbb{R}^n$ and the Haar measure $\dfrac{\boldsymbol{d}x_1}{x_1} \cdots \dfrac{\boldsymbol{d}x_n}{x_n}$. The group Fourier transformation on $\mathbb{R}^n_+$ coincides with the Mellin transformation and its inverse

$$\mathcal{M}_\beta\psi(\xi) := \int\limits_{\mathbb{R}^n_+} t^{i\xi-\beta}\psi(t)\,\frac{dt_1}{t_1}\cdots\frac{dt_n}{t_n}, \quad \xi\in\mathbb{R}^n,$$

$$\mathcal{M}_\beta^{-1}\varphi(t) := \frac{1}{(2\pi)^n}\int\limits_{\mathbb{R}^n_+} t^{\beta-i\xi}\varphi(\xi)\,\frac{d\xi_1}{\xi_1}\cdots\frac{d\xi_n}{\xi_n}, \qquad t\in\mathbb{R}^n_+,$$

$$\beta=(\beta_1,\dots,\beta_n),\quad \beta_j>-1,\quad j=1,\dots,n.$$

Convolution equation is ($g(\xi)=c_0+\mathcal{M}_\beta k(\xi)$)

$$\mathfrak{M}^0_g\varphi(t)=c_0\varphi(t)+\int\limits_{\mathbb{R}^n_+} k\left(\frac{t_1}{\tau_1},\dots,\frac{t_n}{\tau_n}\right)\varphi(\tau)\,\frac{dt_1}{t_1}\cdots\frac{dt_n}{t_n}=f(t),\quad t\in\mathbb{R}^n_+.$$

The generic differential operators in $\mathbb{R}^n_+$ are

$$\mathfrak{D}_k=t_k\partial_k=\mathbf{W}^0_{-i\xi_k+\beta_k},\qquad k=1,2,\dots,n.$$

Example 3.3 (cf. [3]) The direct product of intervals $I^n=(-1,1)^n$ is a Lie group with the group operation

$$x\circ y=\left(\frac{x_1+y_1}{1+x_1y_1},\dots,\frac{x_1+y_1}{1+x_1y_1}\right),\qquad x,y\in I^n$$

with the dual group $\widehat{I^n}=\mathbb{R}^n$ and the Haar measure $\dfrac{dx_1}{1-x_1^2}\cdots\dfrac{dx_n}{1-x_n^2}$. The group Fourier transformation on I^n and its inverse are

$$(\mathcal{F}_{I^n}v)(\xi):=\int_{I^n}\left(\frac{1+y}{1-y}\right)^{i\xi}v(y)\frac{dy_1}{1-y_1^2}\cdots\frac{dy_n}{1-y_n^2},\qquad \xi\in\mathbb{R}^n,$$

$$(\mathcal{F}_{I^n}^{-1}\psi)(x)=\frac{1}{\pi^n}\int_{-\infty}^{\infty}\left(\frac{1+x}{1-x}\right)^{-i\xi}\psi(\xi/2)d\xi,\quad x\in I^n.$$

Convolution equation is

$$(\mathbf{W}^0_{I^n,g}\varphi)(x)=c_0u(x)+\int_{I^n}k\left(\frac{x_1-y_1}{1-x_1y_1},\dots\frac{x_n-y_n}{1-x_ny_n}\right)v(y)\times$$
$$\times\frac{dy_1}{1-y_1^2}\cdots\frac{dy_n}{1-y_n^2}=h(x),\qquad g(\xi)=(\mathcal{F}_{I^n}k)(\xi),\quad \xi\in\mathbb{R}^n,\quad x\in I^n.$$

Here the generic differential operators are

$$\mathfrak{D}_k=(1-t_k^2)\partial_k=\mathbf{W}^0_{-2i\xi_k},\qquad k=1,2,\dots,n.$$

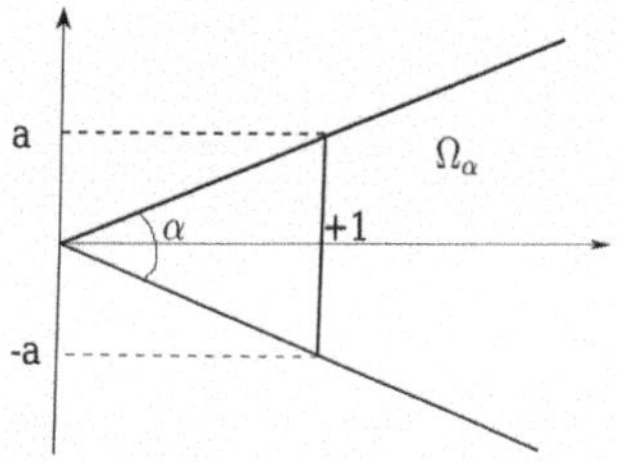

Fig. 1 Angular domain

Example 3.4 The angular domain of magnitude α (or the 2D cone; see Fig. 1)

$$\Omega_\alpha := \{(x_1, x_1x_2) \ : \ x_1 \in \mathbb{R}^+, \ \ |x_2| < a\}$$
$$:= \{(x_1, x_2) \ : \ x_1 \in \mathbb{R}^+, \ \ |x_2| < ax_1\}$$
$$a = 2\cot\frac{\alpha}{2}, \qquad 0 < \alpha < \pi$$

represents a Lie group with respect to the group operation

$$(x_1, x_2) \circ (y_1, y_2) = \left(x_1y_1, ax_1y_1\frac{ax_2y_1 + ax_1y_2}{a^2x_1y_1 + x_2y_2}\right).$$

The neutral element (identity) in the group is $e = (1, 0)$.

The dual group is $\widehat{\Omega_\alpha} = \mathbb{R}^2$ and the Haar measure is $\dfrac{a dx_1 dx_2}{(ax_1)^2 - x_2^2}$. The group Fourier transformation on Ω_α and its inverse are

$$(\mathcal{F}_{\Omega_\alpha}\varphi)(\xi) = \int_{\Omega_\alpha} x_1^{i\xi_1}\left(\frac{ax_1 + x_2}{ax_1 - x_2}\right)^{i\xi_2/2}\frac{\varphi(x_1, x_2)adx_1dx_2}{(ax_1)^2 - x_2^2}, \qquad \xi \in \mathbb{R}^2$$

$$(\mathcal{F}^{-1}_{\Omega_\alpha}\psi)(x) = \frac{1}{(2\pi)^2}\int_{\Omega_\alpha} x_1^{-i\xi_1}\left(\frac{ax_1 + x_2}{ax_1 - x_2}\right)^{-i\xi_2/2}\psi(\xi_1, \xi_2)d\xi_1d\xi_2, \qquad x \in \Omega_\alpha.$$

Convolution equation is

$$(\mathbf{W}^0_{\Omega_\alpha, g}\varphi)(x) = c_0\varphi(x) + \int_{\Omega_\alpha} k\left(\frac{x_1}{y_1}, a^2\frac{x_1}{y_1}\frac{x_2y_1 - x_1y_2}{a^2x_1y_1 - x_2y_2}\right)\varphi(y_1, y_2)\times$$
$$\times\frac{a dx_1 dx_2}{(ax_1)^2 - x_2^2}, \qquad x \in \Omega_\alpha.$$

The generic differential operators are

$$\mathfrak{D}_1\varphi(x) = \partial_1 + x_2\partial_2, \qquad \mathfrak{D}_2\varphi(x) = \frac{a^2x_1^2 - x_2^2}{ax_1}\partial_2.$$

Fig. 2 Star shaped domain

Further examples we describe very shortly. Details will be exposed in the forthcoming paper.

Example 3.5 Star shaped domain $\mathcal{S}^n_\gamma$ and its surface $\mathbb{S}^{n-1}_\gamma = \partial\mathcal{S}^n_\gamma$ (see Fig. 2 for a 2D case), are described with the help of the polar coordinates as follows

$$\mathcal{S}^n_\gamma := \left\{(r\gamma(\theta),\theta)\ :\ 0<r<1,\quad \theta\in\Pi^{n-1}\right\},$$
$$\Pi^{n-1} := (-\pi,\pi)\times(0,\pi)^{n-2},$$
$$\mathbb{S}^{n-1}_\gamma = \partial\mathcal{S}^n_\gamma := \left\{(\gamma(\theta),\theta)\ :\ \theta\in\Pi^{n-1}\right\}.$$

The function $\gamma(\theta)\ :\ (-\pi,\pi)\times(0,\pi)^{n-2}\to(0,a)$, $a<\infty$, be 2π periodic with respect to the variable θ_1 and π-periodic with respect to all other variables $\theta_2,\dots,\theta_{n-1}$.

If we impose only boundedness condition on the derivative of $\gamma(\theta)$, i.e., if $\gamma\in\mathbb{W}^1_\infty(\Pi^{n-1})$, then the boundary hypersurface $\mathbb{S}^n_\gamma = \partial\mathcal{S}^n_\gamma$ is rather rough and might have angular points, edges and even incoming and outgoing peaks. The group binary operations for these Lie groups are defined as follows

$$(r_1\gamma(\theta^1),\theta^1)\circ(r_2\gamma(\theta^2),\theta^2) := (r_1\circ r_2\gamma(\theta^1\circ\theta^2),\theta^1\circ\theta^2),$$
$$(r_j\gamma(\theta^j),\theta^j)\in\mathcal{S}^n_\gamma,$$
$$r_1\circ r_2 := \frac{r_1r_2}{1+2r_1r_2-r_1-r_2},\qquad \theta^1_1\circ\theta^2_1 := \pi^2\frac{\theta^1_1+\theta^2_1}{\pi^2+\theta^1_1\theta^2_1}$$
$$(r_j\gamma(\theta^j),\theta^j)\in\mathcal{S}^n,\quad j=1,2,\quad r_1,r_2\in(0,1),$$
$$\theta^1_1,\theta^1_1\in(-\pi,\pi),$$
$$\theta^1_k\circ\theta^2_k := \pi\frac{\theta^1_k\theta^2_k}{\pi^2+2\theta^1_k\theta^2_k-\theta^1_k-\theta^2_k},$$
$$\theta^1_k,\theta^2_k\in(0,\pi),\quad k=2,\dots,n-1.$$

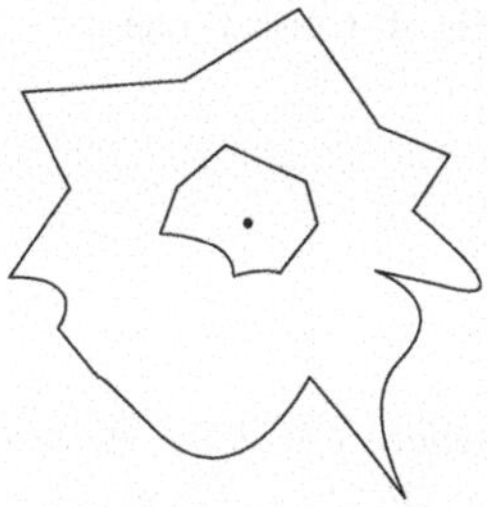

Fig. 3 Star shaped domain with cavity

$$(\gamma(\theta^1), \theta^1) \circ (\gamma(\theta^2), \theta^2) := (\gamma(\theta^1 \circ \theta^2), \theta^1 \circ \theta^2),$$
$$(\gamma(\theta^j), \theta^j) \in \mathbb{S}_\gamma^{n-1}, \quad j = 1, 2.$$

Example 3.6 Star shaped domains with a star shaped cavity, provided they have a common central point O (see Fig. 3 for a 2D case), is described with the help of the polar coordinates as follows:

$$\mathcal{S}_{\omega,\gamma}^n := \big\{(\omega(\theta) + r\gamma(\theta) \ : \ 0 < r < 1, \qquad \theta \in \Pi^{n-1}\big\},$$

where $\Pi^{n-1} := (-\pi, \pi) \times (0, \pi)^{n-2}$, and functions $\gamma(\theta), \omega(\theta) \ : \ \Pi^{n-1} \to (0, a)$, $a < \infty$, are as in Example 3.5.

As in the foregoing Example 3.5, the boundary hypersurface, might be rather rough and have angular points, edges and even incoming and outgoing peaks.

The group binary operation for the star shaped domain with a star shaped cavity $\mathcal{S}_{\omega,\gamma}^n$ is defined as follows

$$(\omega(\theta^1) + r_1\gamma(\theta^1), \theta^1) \circ (\omega(\theta^2) + r_2\gamma(\theta^2), \theta^2)$$
$$:= (\omega(\theta^1 \circ \theta^2) + r_1 \circ r_2\gamma(\theta^1 \circ \theta^2), \theta^1 \circ \theta^2),$$

where the binary operations $r_1 \circ r_2$ and $\theta^1 \circ \theta^2$, $r_1, r_2 \in (0, 1)$, $\theta^m = (\theta_1^m, \ldots, \theta_{n-1}^m) \in \Pi^{n-1}$, $m = 1, 2$, are defined in Example 3.5.

Example 3.7 Semi-infinite $\mathcal{C}_\gamma^{n+1}$, finite $\mathcal{C}_{c,\gamma}^{n+1}$ cones based on star shaped domains and their lateral surfaces $\mathbb{S}\mathcal{C}_\gamma^{n+1}$, $\mathbb{S}\mathcal{C}_{c,\gamma}^{n+1}$

$$\begin{aligned}
\mathcal{C}_\gamma^{n+1} &:= \Big\{(t, r\gamma(t,\theta), \theta) \ : \ r \in (0, 1), \quad t \in \mathbb{R}^+, \quad \theta \in \Pi^{n-1}\Big\} \\
\mathcal{C}_{c,\gamma}^{n+1} &:= \Big\{(t, r\gamma(t,\theta), \theta) \ : \ r \in (0, 1), \quad t \in (0, c), \quad \theta \in \Pi^{n-1}\Big\}, \\
\mathbb{S}\mathcal{C}_\gamma^{n+1} &:= \Big\{(t, \gamma(t,\theta), \theta) \ : \ t \in \mathbb{R}^+, \quad \theta \in \Pi^{n-1}\Big\} \\
\mathbb{S}\mathcal{C}_{c,\gamma}^{n+1} &:= \Big\{(t, \gamma(t,\theta), \theta) \ : \ t \in (0, c), \quad \theta \in \Pi^{n-1}\Big\}
\end{aligned}$$

Fig. 4 Curved cone

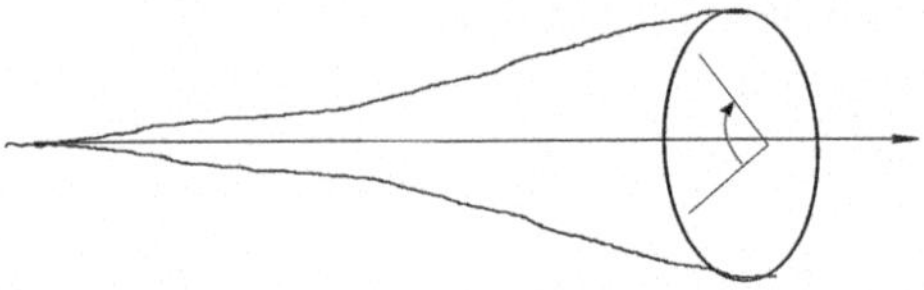

where $\gamma(t,\theta) : \mathbb{R}^+ \times \Pi^{n-1} \to (0,b)$ is a $\mathbb{W}^1_\infty(\mathbb{R}^+ \times \Pi^{n-1})$-function and $\gamma(0,\theta) \equiv 0$, $\Pi^{n-1} := (-\pi,\pi) \times (0,\pi)^{n-2}$. $\gamma(t,\theta)$ is 2π periodic with respect to the variable θ_1 and π-periodic with respect to all other variables $\theta_2,\dots,\theta_{n-1}$.

When $\partial_t\gamma(t,\theta) \equiv at$, $a > 0$, the cone is classical infinite one. If, in addition, $(\partial_t\gamma)(0,\theta) \equiv 0$, the cone is curvilinear and cuspidal, with the cusp at the vertex $t = 0$ (cf. Fig. 4). The binary operations in the above Lie groups are defined as follows The binary operation in the Lie groups $\mathcal{C}^{n+1}_\gamma$, $\mathcal{C}^{n+1}_{c,\gamma}$ and their surfaces $\mathbb{S}\mathcal{C}^{n+1}_\gamma$, $\mathbb{S}\mathcal{C}^{n+1}_{c,\gamma}$ are defined as follows:

$$\begin{aligned}
&(t, r\gamma(t,\theta^1),\theta^1) \circ (\tau, \rho\gamma(\tau,\theta^2),\theta^2)\\
&:=\left(t\tau, r\circ\rho\gamma(t\tau,\theta^1\circ\theta^2),\theta^1\circ\theta^2\right) \quad \text{in} \quad \mathcal{C}_\gamma,\\
&(t, r\gamma(t,\theta^1),\theta^1) \circ (\tau, \rho\gamma(\tau,\theta^2),\theta^2)\\
&:=\left(t\circ\tau, r\circ\rho\gamma(t\circ\tau,\theta^1\circ\theta^2),\theta^1\circ\theta^2\right) \text{ in } \mathcal{C}_{c,\gamma},\\
&(t, \gamma(t,\theta^1),\theta^1) \circ (\tau, \gamma(\tau,\theta^2),\theta^2)\\
&:= \left(t\tau, \gamma(t\tau,\theta^1\circ\theta^2),\theta^1\circ\theta^2\right) \qquad \text{in} \quad \mathbb{S}\mathcal{C}_\gamma,\\
&(t, \gamma(t,\theta^1),\theta^1) \circ (\tau, \gamma(\tau,\theta^2),\theta^2)\\
&:= \left(t\circ\tau, \gamma(t\circ\tau,\theta^1\circ\theta^2),\theta^1\circ\theta^2\right) \quad \text{in} \quad \mathbb{S}\mathcal{C}_{c,\gamma},
\end{aligned}$$

where the binary operations $r\circ\rho$ and $\theta^1\circ\theta^2$, $r,\rho \in (0,1)$, $\theta^m = (\theta^m_1,\dots,\theta^m_{n-1}) \in \Pi^{n-1}$, $m = 1,2$, are defined in Example 3.5 and $t\circ\tau := t\tau$ for $t,\tau \in \mathbb{R}^+$, $t\circ\tau := \dfrac{ct\tau}{2t\tau - ct - c\tau + c^2}$ for $t,\tau \in (0,c)$.

Example 3.8 Infinite, semi-infinite and finite cylinders based on star shaped domains and their lateral surfaces (cf. Fig. 5):

$$\mathfrak{CS}^{n+1}_{\omega,\gamma,\mathcal{I}} := \left\{(t, r\omega(t)\gamma(\theta),\theta) \ : \ r\in[0,1],\ \ t\in\mathcal{I},\quad \theta\in(-\pi,\pi)\times[0,\pi]^{n-2}\right\},$$

$$\mathfrak{CS}^{n}_{\omega,\gamma,\mathcal{I}} := \left\{(t, \omega(t)\gamma(\theta),\theta) \ : \ t\in\mathcal{I},\quad \theta\in(-\pi,\pi)\times[0,\pi]^{n-2}\right\},$$

where if $\mathcal{I} = \mathbb{R}$ we have infinite cylinder, if $\mathcal{I} = \mathbb{R}^+$ we have semi-infinite cylinder and if $\mathcal{I} = (-1,1)$ we have finite cylinder. $\omega : \mathbb{R} \to (0,a)$ is a function (like $\gamma(\theta)$)

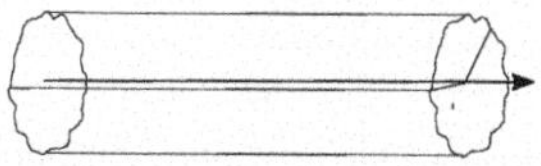

Fig. 5 Semi-infinitr cylinder

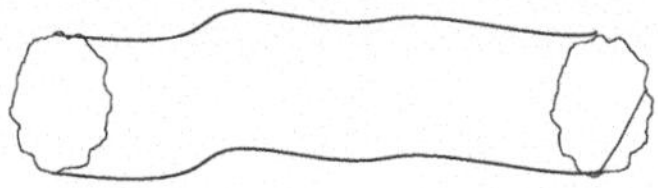

Fig. 6 Curved cylinder (waveguide)

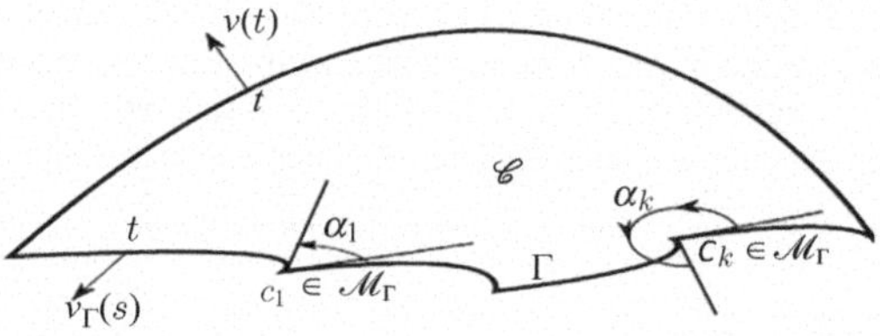

Fig. 7 Hypersurface

with a minimal constraint $\omega \in \mathbb{W}^1_\infty(\mathbb{R})$. The boundary of $\mathfrak{C}\mathcal{S}^{n+1}_{\omega,\gamma,\mathcal{I}}$ and the surface $\mathfrak{C}\mathbb{S}^n_{\omega,\gamma,\mathcal{I}}$ have angular points, edges, incoming and outgoing cusps along the axes t as well as along boundary of their cross sections.

Example 3.9 Curved cylinders (waveguides) and cones (revisited). Consider cylinders based on star shaped domains and curved lateral surfaces (cf. Fig. 6)

$$\mathfrak{C}^+_{\omega_+,\gamma,\beta} := \Big\{(\omega_+(x), r\beta(x)\gamma(\theta), \theta) \ : \ r \in [0,1], \quad x \in \mathbb{R}^+, \quad \theta \in (0,\pi)\Big\},$$

where $\beta : \mathbb{R}^+ \to (0,a)$ is a function (like $\gamma(\theta)$) with a minimal constraint $\beta \in \mathbb{W}^1_\infty(\mathbb{R}^+)$. But we skip the details about the curved cylinder $\mathfrak{C}^+_{\omega_+,\gamma,\beta}$. for simplicity.

A similar description is available also for finite and infinite cylinders and cones (and semi-infinite cones) but, for simplicity, we will consider in details only 3D semi-infinite curved cylinder.

Example 3.10 A hypersurface with the Lipschitz boundary in $\mathbb{R}^n$ as an image of a star-shaped domain $\Omega \subset \mathbb{R}^{n-1}$ (see Fig. 7) is a Lie group as well.

Acknowledgments The second author (R. Duduchava) was supported by the grant of the University of Georgia "Mathematical Analysis and PDE on Lie Groups".

References

1. D. Cardona, R. Duduchava, A. Hendrickx, M. Ruzhansky, Global pseudo-differential operators on the Lie group $G = (-1, 1)^n$. Integral Equations and Operator Theory, 34 pp. (2022), Submitted. https://arxiv.org/abs/2209.09751
2. R. Duduchava, *Integral Equations with Fixed Singularities* (Teubner, Leipzig, 1979)

3. R. Duduchava, Convolution equations on the Lie group G = (-1,1). Georgian Math. J. **30**(5), 683–702 (2023). https://doi.org/10.1515/gmj-2023-2035
4. V. Fischer, M. Ruzhansky, *Quantization on Nilpotent Lie Groups*, Progress in Mathematics, vol. 314 (Birkhauser, 2016), xiii+557pp
5. I. Gohberg, N. Krupnik, *One-Dimensional Linear Singular Integral Equations*, vols. I–II, vol. 53–54 of Operator Theory, Advances and Applications (Birkhäuser Verlag, Basel, 1979)
6. L. Hörmander, Estimates for translation invariant operators in L_p spaces. Acta Math., **104**, 93–140 (1960)
7. S. Igari, Functions of L_p-multipliers. Tohoku Math. J. **21**, 304–320 (1969)
8. I. Simonenko, A new general method of investigating linear operator equations of singular integral equation type. I. (Russian). Izv. Akad. Nauk SSSR Ser. Mat. **29**, 567–586 (1965)

The Finite Hilbert Transform Acting in Rearrangement Invariant Spaces on (−1, 1)

Guillermo P. Curbera, Susumu Okada, and Werner J. Ricker

Abstract We present recent results in the study of the finite Hilbert transform in rearrangement invariant spaces on $(-1, 1)$.

Keywords Finite Hilbert transform · Airfoil equation · Rearrangement invariant space · Spectrum · Zygmund space $L\log L$

2020 Mathematics Subject Classification Primary 44A15, 46E30; Secondary 47A53, 47B34, 28B05

This note gives account of an invited talk delivered by the first author at the Tbilisi Analysis & PDE Workshop (TAPDE 2023) held in Tbilisi, Georgia, from August 30 to September 2, 2023.

For $f \in L^1 := L^1(-1, 1)$ its finite Hilbert transform $T(f)$ is the p.v.-integral

$$T(f)(t) = \lim_{\varepsilon \to 0^+} \frac{1}{\pi} \left(\int_{-1}^{t-\varepsilon} + \int_{t+\varepsilon}^{1} \right) \frac{f(x)}{x - t}\, dx, \tag{1}$$

G. P. Curbera (✉)
Facultad de Matemáticas, Universidad de Sevilla, Sevilla, Spain
e-mail: curbera@us.es

S. Okada
Kambah, ACT, Australia

W. J. Ricker
Math.–Geogr. Fakultät, Katholische Universität Eichstätt–Ingolstadt, Eichstätt, Germany
e-mail: werner.ricker@ku.de

R. Duduchava et al. (eds.), *Tbilisi Analysis and PDE Seminar*,
Research Perspectives Ghent Analysis and PDE Center 7,
https://doi.org/10.1007/978-3-031-62894-8_6

which exists for a.e. $t \in (-1, 1)$ and is a measurable function. We will abbreviate it as FHT. Its study is related to the solution of the airfoil equation: for g a given function, find all functions f which satisfy

$$g(t) = \text{p.v.}\frac{1}{\pi}\int_{-1}^{1}\frac{f(x)}{x-t}\,dx, \quad \text{a.e. } t \in (-1, 1). \tag{2}$$

Important references for the study of the FHT and related singular integral operators are [7, 9–14, 19, 20, 22].

The setting of this note is the family of rearrangement invariant (r.i.) spaces on $(-1, 1)$, which includes classical spaces from analysis such as the Lorentz $L^{p,q}$ spaces, Orlicz L^{φ} spaces, Marcinkiewicz M_{φ} spaces, Lorentz Λ_{φ} spaces, and the Zygmund $L^p(\log \mathrm{L})^{\alpha}$ spaces. Standard references for r.i. spaces are [1, 15, 16].

Two formulae are fundamental for the investigation of the FHT:

(a) The Parseval formula holds for a pair of functions $f, g \in L^1$ if

$$\int_{-1}^{1} fT(g) = -\int_{-1}^{1} gT(f). \tag{3}$$

(b) The Poincaré-Bertrand formula holds for a pair of functions $f, g \in L^1$ if

$$T(gT(f) + fT(g)) = T(f)T(g) - fg, \quad \text{a.e. on } (-1, 1). \tag{4}$$

Parseval's formula (3) holds for $f \in L^p$ and $g \in L^{p'}$ with $1 < p < \infty$ and $1/p + 1/p' = 1$, [22, §4.3 (2)]. The Poincaré-Bertrand formula (4) was proved by Tricomi for $f \in L^p$ and $g \in L^q$ with $1/p + 1/q < 1$, [22, §4.3 (4)]). In 1977, Love established (4) for $f \in L^p$ and $g \in L^{p'}$, [17, Corollary].

The weight function $w(x) := \sqrt{1-x^2}$, $x \in (-1, 1)$, plays a fundamental role in the theory of the FHT since $T(1/w) = 0$, [22, §4.3 (7)], and because it determines the kernel of T, [22, §4.3 (14)], namely,

$$f \in \bigcup_{1<p<\infty} L^p \text{ satisfies } T(f) = 0 \iff f = C/w \text{ for some } C \in \mathbb{C}. \tag{5}$$

Note that the function $1/w$ belongs to the Marcinkiewicz space $L^{2,\infty}$, which is, in fact, the smallest r.i. space containing $1/w$.

We briefly recall the tools used in the L^p-theory of the FHT and the airfoil equation. A celebrated theorem of M. Riesz states, for each $1 < p < \infty$, that the Hilbert transform H maps $L^p(\mathbb{R})$ continuously into itself, [1, Theorem III.4.9(a)]. It follows that $T : L^p \to L^p$ continuously. We denote this operator by T_p.

For each $f \in L^1$, the function $\widehat{T}(f) \in L^0$ is defined by

$$\widehat{T}(f) := -\frac{1}{w} T(fw). \tag{6}$$

For $1 < p < 2$ the restriction of $\widehat{T}$ to L^p defines a continuous operator $\widehat{T}_p \colon L^p \to L^p$.

For $f \in L^1$ satisfying $f/w \in L^1$, the function $\check{T}(f) \in L^0$ is defined by

$$\check{T}(f) := -w\, T\left(\frac{f}{w}\right). \tag{7}$$

For $2 < p < \infty$ the restriction of $\check{T}$ to L^p defines a continuous operator $\check{T}_p \colon L^p \to L^p$.

The operators $\widehat{T}$ and $\check{T}$ provide an understanding of the action of T on the spaces L^p whenever $1 < p < 2$ and $2 < p < \infty$, respectively, as well as obtaining an inversion formula for the airfoil equation (2) within L^p; see [21, §2].

A result of Boyd allows an extension of M. Riesz's result to a larger class of r.i. spaces. Given any r.i. space Y over $\mathbb{R}$, Boyd associated two indices, $\underline{\alpha}_Y$ and $\overline{\alpha}_Y$, to Y which satisfy $0 \leq \underline{\alpha}_Y \leq \overline{\alpha}_Y \leq 1$, and proved that H acts continuously on Y if and only if these indices are non-trivial, that is, $0 < \underline{\alpha}_Y \leq \overline{\alpha}_Y < 1$, [1, Theorem III.5.18]. The indices $\underline{\alpha}_Y$ and $\overline{\alpha}_Y$ are called the lower and upper Boyd indices of Y, resp. Regarding the FHT, the analogous characterization as above is valid: T acts continuously on a r.i. space X on $(-1, 1)$ if and only if X has non-trivial Boyd indices, [15, pp. 170–171]. To indicate that $T \colon X \to X$ continuously we write T_X.

The Parseval formula (3) and the Poincaré-Bertrand formula (4) were extended to any pair of functions $f \in X$ and $g \in X'$ (the associate space of X) and all r.i. spaces X on $(-1, 1)$ with non-trivial Boyd indices, [2, Proposition 3.1]. The validity of both formulae can be further extended, [4, Theorems 3.1 and 3.2].

Theorem 1 *Let $f \in L^1$ and $g \in L\log L$ satisfy $fT(g\chi_A) \in L^1$, for every set $A \in \mathcal{B}$. Then the Parseval formula* (3) *and the Poincaré-Bertrand formula* (4) *are valid.*

Note, for r.i. spaces X on $(-1, 1)$, that

$$\bigcup_{0<\underline{\alpha}_X \leq \overline{\alpha}_X < 1} X = \bigcup_{1<p<\infty} L^p \subsetneq L\log L.$$

The extended version of the Poincaré-Bertrand formula given in Theorem 1 allows an extension of (5) when T acts in $L\log L$, [4, Theorem 3.4].

Theorem 2 *Let $f \in L\log L$. Then $T(f) = 0$ in L^1 if and only if $f = C/w$, for some constant $C \in \mathbb{C}$.*

These results above describing the action of T on L^p for $1 < p < 2$ and $2 < p < \infty$ can be extended to the larger classes of r.i. spaces X satisfying $1/2 < \underline{\alpha}_X \leq \overline{\alpha}_X < 1$ and $0 < \underline{\alpha}_X \leq \overline{\alpha}_X < 1/2$, respectively, [2, Theorems 3.2 and 3.3].

Theorem 3 *Let X be a r.i. space satisfying $1/2 < \underline{\alpha}_X \leq \overline{\alpha}_X < 1$.*

(i) $\operatorname{Ker}(T_X) = \operatorname{span}\{1/w\}$.
(ii) *The linear operator $\widehat{T}$ maps X continuously into X. The resulting operator $\widehat{T}_X : X \to X$ satisfies $T_X \widehat{T}_X = I_X$. Moreover,*

$$\int_{-1}^{1} \widehat{T}_X(f)(x)\, dx = 0, \quad f \in X.$$

(iii) *The operator $T_X : X \to X$ is surjective.*
(iv) *The identity $\widehat{T}_X T_X = I_X - P_X$ holds, with P_X the projection given by*

$$P_X(f) := \left(\frac{1}{\pi}\int_{-1}^{1} f(t)\, dt\right)\frac{1}{w}, \quad f \in X.$$

(v) *The operator $\widehat{T}_X$ is an isomorphism onto its range $R(\widehat{T}_X)$. Moreover,*

$$R(\widehat{T}_X) = \left\{ f \in X : \int_{-1}^{1} f(x)\, dx = 0 \right\}.$$

(vi) *The following direct sum decomposition holds: $X = R(\widehat{T}_X) \oplus \operatorname{span}\{1/w\}$.*

Theorem 4 *Let X be a r.i. space satisfying $0 < \underline{\alpha}_X \leq \overline{\alpha}_X < 1/2$.*

(i) *The operator $T_X : X \to X$ is injective.*
(ii) *The operator $\check{T}_X$ is continuous from X into X and satisfies $\check{T}_X T_X = I_X$.*
(iii) *The identity $T_X \check{T}_X = I_X - Q_X$ holds, with Q_X the projection given by*

$$Q_X(f) := \left(\frac{1}{\pi}\int_{-1}^{1} \frac{f(x)}{w(x)}\, dx\right)\chi_{(-1,1)}, \quad f \in X.$$

(iv) *The range of T_X is the closed subspace of X given by*

$$R(T_X) = \left\{ f \in X : \int_{-1}^{1} \frac{f(x)}{w(x)}\, dx = 0 \right\} = \operatorname{Ker}(Q_X).$$

Moreover, $\check{T}_X$ is an isomorphism from $R(T_X)$ onto X.
(v) *The following direct sum decomposition holds: $X = R(T_X) \oplus \operatorname{span}\{\chi_{(-1,1)}\}$.*

Theorems 3 and 4 lead to the following general result on the inversion of the airfoil equation, [2, Corollary 3.5], which is an extension of the L^p-case.

Corollary 5 *Let X be a r.i. space.*

(i) *Suppose that $1/2 < \underline{\alpha}_X \leq \overline{\alpha}_X < 1$ and $g \in X$ is fixed. Then all solutions $f \in X$ of the airfoil equation* (2) *are given, with $C \in \mathbb{C}$ arbitrary, by*

$$f = \frac{-1}{w} T_X(wg) + \frac{C}{w}.$$

(ii) *Suppose that $0 < \underline{\alpha}_X \leq \overline{\alpha}_X < 1/2$ and $g \in X$ satisfies $\int_{-1}^{1} \frac{g(x)}{w(x)} dx = 0$. Then there is a unique solution $f \in X$ of the airfoil equation* (2)*, namely*

$$f = -w\, T_X\Big(\frac{g}{w}\Big).$$

The previous result does not apply to r.i. spaces X satisfying $\underline{\alpha}_X \leq 1/2 \leq \overline{\alpha}_X$, which includes the Lorentz spaces $L^{2,q}$ with $1 \leq q \leq \infty$ and, in particular, L^2.

The spectrum of $T_p \colon L^p \to L^p$, for $1 < p < \infty$, was completely identified by Widom in 1960, who also described its fine spectra, that is, the point spectrum, continuous spectrum and residual spectrum, [23]; see also [12, §13.6].

For $1 < p < \infty$, consider the subset of $\mathbb{C}$ given by

$$\mathcal{R}_p := \{\pm 1\} \cup \left\{\lambda \in \mathbb{C} : \frac{1}{2\pi}\left|\arg\left(\frac{1+\lambda}{1-\lambda}\right)\right| \leq \left|\frac{1}{2} - \frac{1}{p}\right|\right\}.$$

The following result is due to Widom, [23, Remarks 1(2) and 2, pp.156–157], where the FHT used differs from (1) by a factor of $1/i$.

Theorem 6 *Let $1 < p < \infty$. The operator $T_p \colon L^p \to L^p$ has spectrum $\sigma(T_p) = \mathcal{R}_p$.*

Regarding the fine spectra of T_p the following identifications hold.

(i) *Let $1 < p < 2$. Then $\sigma_{\mathrm{pt}}(T_p) = \mathrm{int}(\mathcal{R}_p)$, $\sigma_{\mathrm{r}}(T_p) = \emptyset$ and $\sigma_{\mathrm{c}}(T_p) = \partial\mathcal{R}_p$.*
(ii) *Let $p = 2$. Then $\sigma_{\mathrm{pt}}(T_2) = \emptyset$, $\sigma_{\mathrm{r}}(T_2) = \emptyset$ and $\sigma_{\mathrm{c}}(T_2) = \partial\mathcal{R}_2 = \mathcal{R}_2$.*
(iii) *Let $2 < p < \infty$. Then $\sigma_{\mathrm{pt}}(T_p) = \emptyset$, $\sigma_{\mathrm{r}}(T_p) = \mathrm{int}(\mathcal{R}_p)$ and $\sigma_{\mathrm{c}}(T_p) = \partial\mathcal{R}_p$.*

Jörgens, in 1970, identified the set of all possible eigenfunctions for T when T acts over $\bigcup_{1<p<\infty} L^p$; see [12, Theorem 13.9].

Returning to the spectrum of T_X, the following result identifies the role played by the location of the point 0 within $\sigma(T_X)$, [3, Proposition 5.1].

Proposition 7 *Let X be a separable r.i. space with non-trivial Boyd indices. Precisely one of the following three mutually exclusive alternatives holds.*

(i) *The following conditions are equivalent to $0 \in \sigma_{\mathrm{pt}}(T_X)$.*

(a) $\sigma_{\mathrm{pt}}(T_X) \neq \emptyset$.
(b) $|x|^{-1/2} \in X$.

(c) $L^{2,\infty} \subseteq X$.
(d) $(-1, 1) \subseteq \sigma_{\mathrm{pt}}(T_X)$.

(ii) *The following conditions are equivalent to* $0 \in \sigma_{\mathrm{r}}(T_X)$.

(a) $\sigma_{\mathrm{r}}(T_X) \neq \emptyset$.
(b) $|x|^{-1/2} \in X'$.
(c) $X \subseteq L^{2,1}$.
(d) $(-1, 1) \subseteq \sigma_{\mathrm{r}}(T_X)$.

(iii) *The following conditions are equivalent to* $0 \in \sigma_{\mathrm{c}}(T_X)$.

(a) $|x|^{-1/2}$ *belongs to neither* X *nor to* X'.
(b) $\sigma(T_X) = \sigma_{\mathrm{c}}(T_X) \supseteq [-1, 1]$.

Given any r.i. space X with non-trivial Boyd indices, define $p_X \in (1, \infty)$ by

$$p_X := \inf\Big\{p \in (1, \infty) : L^{p,\infty} \subseteq X\Big\}. \tag{8}$$

It allows a characterization of the point spectrum of T_X, [3, Proposition 3.3].

Proposition 8 *Let* X *be any r.i. space with non-trivial Boyd indices.*

(i) *If* $p_X > 2$ *(attained or not) or,* $p_X = 2$ *(not attained), then* $\sigma_{\mathrm{pt}}(T_X) = \emptyset$.
(ii) *If* $p_X \leq 2$ *with* p_X *attained, then* $\sigma_{\mathrm{pt}}(T_X) = \mathcal{R}_{p_X} \setminus \{\pm 1\}$.
(iii) *If* $p_X < 2$ *with* p_X *not attained, then* $\sigma_{\mathrm{pt}}(T_X) = \mathrm{int}(\mathcal{R}_{p_X})$.

This result, together with Boyd's interpolation theorem, [16, Theorem 2.b.11], allows a full description of the fine spectra of $T_{p,r} := T : L^{p,r} \to L^{p,r}$, [3, §4].

$L^{p,r}$	$\sigma(T_{p,r}) = \mathcal{R}_p$	$\sigma_{\mathrm{pt}}(T_{p,r})$	$\sigma_{\mathrm{r}}(T_{p,r})$	$\sigma_{\mathrm{c}}(T_{p,r})$
$1 < p < 2$	$1 \leq r < \infty$	$\mathrm{int}(\mathcal{R}_p)$	$\emptyset$	$\partial\mathcal{R}_p$
$2 < p < \infty$	$r = 1$	$\emptyset$	$\mathcal{R}_p \setminus \{\pm 1\}$	$\{\pm 1\}$
	$1 < r < \infty$	$\emptyset$	$\mathrm{int}(\mathcal{R}_p)$	$\partial\mathcal{R}_p$
$p = 2$	$r = 1$	$\emptyset$	$(-1, 1)$	$\{\pm 1\}$
	$1 < r < \infty$	$\emptyset$	$\emptyset$	$[-1, 1]$

Let X be a r.i. space with non-trivial Boyd indices. Define $q_X \in (1, \infty)$ by

$$q_X := \sup\Big\{q \in (1, \infty) : X \subseteq L^{q,1}\Big\}.$$

Note that $q_X \leq p_X$. The following result provides a description of parts of the fine spectra of T_X for a large class of r.i. spaces X.

Proposition 9 *Let X be a separable r.i. space with non-trivial Boyd indices.*

(i) *If both* $p_X < 2$ *and* p_X *is not attained, then*

$$\sigma_{\mathrm{pt}}(T_X) = \mathrm{int}(\mathcal{R}_{p_X});\ \sigma_{\mathrm{r}}(T_X) = \emptyset,$$

whereas if both $p_X \leq 2$ *and* p_X *is attained, then*

$$\sigma_{\mathrm{pt}}(T_X) = \mathcal{R}_{p_X} \setminus \{\pm 1\};\ \sigma_{\mathrm{r}}(T_X) = \emptyset.$$

(ii) *If both* $q_X > 2$ *and* q_X *is not attained, then*

$$\sigma_{\mathrm{pt}}(T_X) = \emptyset;\ \sigma_{\mathrm{r}}(T_X) = \mathrm{int}(\mathcal{R}_{q_X}).$$

whereas if both $q_X \geq 2$ *and* q_X *is attained, then*

$$\sigma_{\mathrm{pt}}(T_X) = \emptyset;\ \sigma_{\mathrm{r}}(T_X) = \mathcal{R}_{q_X} \setminus \{\pm 1\}.$$

(iii) *Suppose that* $q_X \leq 2 \leq p_X$ *and, for those cases when either* $p_X = 2$ *or* $q_X = 2$ *occur, that they are not attained. Then*

$$\sigma_{\mathrm{pt}}(T_X) = \sigma_{\mathrm{r}}(T_X) = \emptyset;\ \sigma_{\mathrm{c}}(T_X) = \sigma(T_X).$$

Identifying $\sigma_{\mathrm{c}}(T_X)$ encounters, in general, serious difficulties. Using an interpolation argument (via Boyd's theorem) it can be shown, [3, Proposition 7.1], that

$$\sigma(T_X) \subseteq \mathcal{R}_{1/\underline{\alpha}_X} \cup \mathcal{R}_{1/\overline{\alpha}_X}, \tag{9}$$

whenever X is separable with $0 < \underline{\alpha}_X \leq \overline{\alpha}_X < 1$. The containment (9), together with earlier results, allows a complete identification of the fine spectra of T_X whenever $0 < \underline{\alpha}_X = \overline{\alpha}_X < 1$, in which case $p_X = q_X = 1/\underline{\alpha}_X = 1/\overline{\alpha}_X$, [3, Theorem 7.2]. The issue of whether the indices p_X, q_X are attained or not attained is indicated in the table by a. or n.a., respectively.

$\sigma(T_X) = \mathcal{R}_{p_X}$	$a./n.a.$	$\sigma_{\mathrm{pt}}(T_X)$	$\sigma_{\mathrm{r}}(T_X)$	$\sigma_{\mathrm{c}}(T_X)$
$p_X < 2$	$n.a.$	$\mathrm{int}(\mathcal{R}_{p_X})$	$\emptyset$	$\partial\mathcal{R}_{p_X}$
	$a.$	$\mathcal{R}_{p_X} \setminus \{\pm 1\}$	$\emptyset$	$\{\pm 1\}$
$p_X > 2$	$n.a.$	$\emptyset$	$\mathrm{int}(\mathcal{R}_{p_X})$	$\partial\mathcal{R}_{p_X}$
	$a.$	$\emptyset$	$\mathcal{R}_{p_X} \setminus \{\pm 1\}$	$\{\pm 1\}$
$p_X = 2$	$p_X a.$	$(-1, 1)$	$\emptyset$	$\{\pm 1\}$
	$q_X a.$	$\emptyset$	$(-1, 1)$	$\{\pm 1\}$
	$p_X, q_X n.a.$	$\emptyset$	$\emptyset$	$\mathcal{R}_2 = [-1, 1]$

We now consider T acting in the classical Zygmund space $L\log L$. Given $\alpha > 0$ the space $L(\log L)^\alpha$ consists of all measurable functions f on $(-1, 1)$ satisfying

$$\int_{-1}^{1} |f(x)|(\log(2+|f(x)|)^\alpha \, dx < \infty.$$

The space $L\log L$, corresponding to $\alpha = 1$, is a r.i. space close to L^1 in the sense that $L^p \subseteq L\log L$ for all $1 < p < \infty$, which implies that $X \subseteq L\log L$ for all r.i. spaces X with non-trivial Boyd indices. The following inclusions hold:

$$L^p \subseteq L(\log L)^\beta \subseteq L(\log L)^\alpha \subseteq L\log L \subseteq L^1, \quad 1 < p, \ 1 < \alpha < \beta.$$

It turns out that $T_{\log} := T : L\log L \to L^1$ is a continuous operator, [4, Theorem 2.1].

Note that Theorems 1 and 2 imply the next result.

Corollary 10 *The Parseval formula*

$$\int_{-1}^{1} fT(g) = -\int_{-1}^{1} gT(f), \quad f \in L^\infty, \ g \in L\log L,$$

holds, as does the Poincaré-Bertrand formula

$$T(gT(f) + fT(g)) = (T(f))(T(g)) - fg, \quad f \in L^\infty, \ g \in L\log L.$$

Moreover, $\mathrm{Ker}(T_{\log}) = \mathrm{span}\{1/w\}$.

In order to study the inversion of the FHT on $L\log L$, the operator $\widehat{T}$ defined in (6) plays a central role. Extrapolation results enter in the study of $T_{\log}$. In particular, the following theorem of Yano is important, [24, Theorem XII.(4.41)].

Theorem 11 *Let* $1 < p_0 < \infty$ *and* S *be a linear operator that maps* L^p *continuously into* L^p *for all* $1 < p < p_0$ *and such that there exist constants* $C > 0$ *and* $1 < \alpha \leq p_0$ *satisfying*

$$\|S\|_{L^p \to L^p} \leq \frac{C}{p-1}, \quad p \in (1, \alpha).$$

Then S *can be extended to* $L\log L$ *with* $S : L\log L \to L^1$ *a continuous operator.*

The following related result, [8, Theorem 5.1], is also required.

Theorem 12 *Let* p_0, S, C *and* α *satisfy the conditions in Yano's theorem. Suppose, for some* $\gamma \geq 0$, *that* $S : L(\log L)^\gamma \to L^1$ *is continuous. Then, continuously, also*

$$S : L(\log L)^{\gamma+\beta} \to L(\log L)^\beta, \quad \forall \beta > 0.$$

Pichorides' calculation of the norm of the Hilbert transform H on $L^p(\mathbb{R})$, together with a result of McLean and Elliott showing that $\|T\|_{L^p\to L^p} = \|H\|_{L^p(\mathbb{R})\to L^p(\mathbb{R})}$, [18, Theorem 3.4], imply that $\|T\|_{L^p\to L^p} = \tan(\pi/(2p)) \le \frac{3}{p-1}$, for $1 < p < 2$, Combined with Theorem 12, this yields the following result, [4, Proposition 4.4].

Proposition 13 *For every $\beta \ge 0$, the finite Hilbert transform T acts continuously*

$$T\colon L(\log L)^{1+\beta} \to L(\log L)^{\beta}.$$

The proof of the corresponding result for the operator $\widehat{T}$ defined in (6) follows a similar strategy. However, it requires not only that $\widehat{T}_p\colon L^p \to L^p$ continuously, for $1 < p < 2$, (a special case of a result of Khvedelidze [13]), but also an explicit upper bound on the operator norms $\|\widehat{T}\|_{L^p\to L^p}$ for p near 1, which is not given in [10, 19]. This upper bound is established via some technical facts, [4, Lemmas 4.6, 4.7, 4.8 and 6.1], which then lead to the following result, [4, Proposition 4.5].

Proposition 14 *For each $\beta \ge 0$, the operator*

$$\widehat{T}\colon L(\log L)^{1+\beta} \to L(\log L)^{\beta}$$

continuously. In particular, $\widehat{T}\colon L\log L \to L^1$ is continuous.

For $T_{\log}\colon L\log L \to L^1$, the result corresponding to Theorems 3 and 4 (for a class of r.i. spaces) is the following one, [4, Theorems 3.4 and 4.10].

Theorem 15 *The following assertions are valid.*

(i) Ker $(T_{\log}) = \operatorname{span}\{1/w\}$.
(ii) *Let $g \in L\log L$. Then $\widehat{T}(g) \in L^1$ and $T(\widehat{T}(g)) = g$. Moreover,*

$$\int_{-1}^{1} \widehat{T}(g)(x)\, dx = 0.$$

(iii) *The continuous, linear operator $P\colon L\log L \to L\log L$ given by*

$$P(f)(x) := \left(\frac{1}{\pi}\int_{-1}^{1} f(t)\, dt\right)\frac{1}{w}, \qquad f \in L\log L,$$

is a rank-1 projection. Moreover, $TP\colon L\log L \to L^1$ is the zero operator.
(iv) *For each $f \in L\log L$ it is the case that*

$$f - P(f) = \widehat{T}(T(f)).$$

Moreover, $\widehat{T}T\colon L\log L \to L\log L$.

Theorem 15 yields a description of the range space of $T_{\log}$, [4, Corollary 4.11].

Corollary 16 *A function $g \in L^1$ belongs to the range space $T(L\log L)$ of $T_{\log}$ if and only if it satisfies both $\widehat{T}(g) \in L\log L$ and $T(\widehat{T}(g)) = g$. That is,*

$$T(L\log L) = \left\{g \in L^1 : \widehat{T}(g) \in L\log L,\ T(\widehat{T}(g)) = g\right\}.$$

The information available in Theorem 15 and Corollary 16 leads to an inversion formula for the operator $T_{\log}\colon L\log L \to L^1$, [4, Theorem 4.14].

Theorem 17 *Let g belong to the range of $T_{\log}\colon L\log L \to L^1$. All solutions $f \in L\log L$ of the airfoil equation* (2) *are of the form*

$$f = \frac{-1}{w}\,T(wg) + \frac{C}{w},$$

with $C \in \mathbb{C}$ arbitrary. In particular, this applies to every $g \in L(\log L)^2 \subseteq T(L\log L)$.

For a more extensive survey of the FHT we refer to [6].

Acknowledgments The first author acknowledges support from PID2021-124332NB-C21, FQM-262.

References

1. C. Bennett, R. Sharpley, *Interpolation of Operators* (Academic Press, Boston, 1988)
2. G.P. Curbera, S. Okada, W.J. Ricker, Inversion and extension of the finite Hilbert transform on $(-1, 1)$. Ann. Mat. Pura Appl. (4) **198**, 1835–1860 (2019)
3. G.P. Curbera, S. Okada, W.J. Ricker, Fine spectra of the finite Hilbert transform in function spaces. Adv. Math. **380**, 107597 (29 pp.) (2021)
4. G.P. Curbera, S. Okada, W.J. Ricker, The finite Hilbert transform acting in the Zygmund space $L\log L$. Ann. Sc. Norm. Super. Pisa Cl. Sci., to appear. https://doi.org/10.2422/2036-2145.202203_003 (see also the arXiv)
5. G.P. Curbera, S. Okada, W.J. Ricker, Measure theoretic aspects of the finite Hilbert transform. Math. Nach., to appear
6. G.P. Curbera, S. Okada, W.J. Ricker, The finite Hilbert transform acting on $(-1, 1)$, In *Proceedings of the International Conference on Operator Theory held in Timisoara, Romania, July 2022*
7. R. Duduchava, Integral equations in convolution with discontinuous presymbols. In: *Singular Integral Equations with Fixed Singularities, and Their Applications to Some Problems of Mechanics* (Teubner Verlagsgesellschaft, Leipzig, 1979)
8. D.E. Edmunds, M. Krbec, Variations on Yano's extrapolation theorem. Rev. Mat. Comput. **111**, 111–118 (2005)
9. F.D. Gakhov, *Boundary Value Problems* (Dover Publications, New York, 1990)
10. I. Gohberg, N. Krupnik, *One-Dimensional Linear Singular Integral Operators Vol. I. Introduction*. Operator Theory Advances and Applications, vol. 53 (Birkhäuser, Berlin, 1992)
11. I. Gohberg, N. Krupnik, *One-Dimensional Linear Singular Integral Operators Vol. II. General Theory and Applications*. Operator Theory Advances and Applications, vol. 54 (Birkhäuser, Berlin, 1992)

12. K. Jörgens, *Linear Integral Operators* (Pitman, Boston, 1982)
13. B.V. Khvedelidze, Linear discontinuous boundary problems in the theory of functions, singular integral equations and some of their applications. Akad. Nauk. Gruzin. SSR **23**, 3–158 (1956) (Russian)
14. F.W. King, *Hilbert Transforms Vol. I* (Cambridge University Press, Cambridge, 2009)
15. S.G. Krein, Ju.I. Petunin, E.M. Semenov, *Interpolation of Linear Operators* (American Mathematical Society, Providence, 1982)
16. J. Lindenstrauss, L. Tzafriri, *Classical Banach Spaces Vol. II* (Springer, Berlin, 1979)
17. E.R. Love, Repeated singular integrals. J. Lond. Math. Soc. (2) **15**, 99–102 (1977)
18. W. McLean, D. Elliott, On the p-norm of the truncated Hilbert transform. Bull. Austral. Math. Soc. **38**, 413–420 (1988)
19. S. Mikhlin, S. Prössdorf, *Singular Integral Operators* (Springer, Berlin, 1986)
20. N.I. Muskhelishvili, *Singular Integral Equations* (Wolters-Noordhoff Pub., Groningen, 1967)
21. S. Okada, D. Elliott, The finite Hilbert transform in $\mathcal{L}^2$. Math. Nachr. **153**, 43–56 (1991)
22. F.G. Tricomi, *Integral Equations* (Interscience, New York, 1957)
23. H. Widom, Singular integral equations in L_p. Trans. Amer. Math. Soc. **97**, 131–160 (1960)
24. A. Zygmund, *Trigonometric Series* (Cambridge University Press, Cambridge, 1959)

The Banach Gelfand Triple and Its Role in Classical Fourier Analysis and Operator Theory

Hans G. Feichtinger

Abstract This short notes tries to provide a summary of a program which started almost 45 years ago with the discovery of "*Feichtinger's Algebra* (then called a *new Segal algebra* $\big(S_0(G), \|\cdot\|_{S_0}\big)$), which was part of the habilitation thesis of the author. Meanwhile there is plenty of evidence that this Banach algebra of test functions, together with the Hilbert spaces L^2 and it's dual space, the space of *mild distributions* are extremely versatile and useful tools for many basic questions in Fourier Analysis. The many natural (in fact functorial) properties of these spaces turn them into a natural alternative to the usual Gelfand triple over $\mathbb{R}^d$, namely $(\mathcal{S}, L^2, \mathcal{S}')(\mathbb{R}^d)$. However, since this are three Banach spaces they are forming a *Banach Gelfand Triple*. Although it is well defined for any locally compact Abelian (LCA) group, we will restrict our attention to the Euclidean case, in order to avoid cumbersome notations. Overall the Banach Gelfand Triple $(S_0, L^2, S_0')(\mathbb{R}^d)$ plays a crucial role in a novel approach to Fourier and Time-Frequency Analysis, as developed by the author under the name of *Conceptual Harmonic Analysis*. It is supposed to provide a mathematically sound basis for all the involved steak-holders, meaning mathematicians who care about applications, engineers or physicists who want to turn their heuristic considerations into concise mathematical formulations. At a technical level only basic principles from functional analysis are required.

Keywords Banach Gelfand Triple · Mild distributions · Feichtinger's algebra · Conceptual Harmonic Analysis

2010 Mathematics Subject Classification 43-02, 46-01, 42B10, 42A38, 46E30, 46F05, 47A70

H. G. Feichtinger (✉)
Faculty of Mathematics, University Vienna, Vienna, Austria

Acoustic Research Institute, OEAW, Vienna, Austria
e-mail: hans.feichtinger@univie.ac.at

R. Duduchava et al. (eds.), *Tbilisi Analysis and PDE Seminar*,
Research Perspectives Ghent Analysis and PDE Center 7,
https://doi.org/10.1007/978-3-031-62894-8_7

1 Introduction

Let us start with the definition of $\left(S_0(\mathbb{R}^d), \|\cdot\|_{S_0}\right)$ via the Short-Time Fourier Transform (STFT), which will be denoted by $V_g(f)$. The STFT of a *signal* $f \in \boldsymbol{L}^2(\mathbb{R}^d)$ with respect to the *window* $g \in \boldsymbol{L}^2(\mathbb{R}^d)$, is given by

$$V_g(f)(t,s) = \int_{\mathbb{R}^d} f(y)g(y-x)e^{-2\pi i s y}dy, \quad t,s \in \mathbb{R}^d, \tag{1.1}$$

i.e. as the Fourier transform of the localized version (near t, if g is symmetric and centered around the origin) of f. If the window is a Schwartz function or even better the Gaussian function, given as $g_0(t) = exp(-\pi|t|^2)$ then f can be much more general, even a tempered distribution, and $V_{g_0}(f)$ is defined via the duality with the Schwartz space $\boldsymbol{S}(\mathbb{R}^d)$ of rapidly decreasing functions. Then membership of f in $S_0(\mathbb{R}^d)$ can be described by the finiteness of the $\boldsymbol{L}^1$-norm of $V_{g_0}(f)$, and we put

$$\|f\|_{S_0(\mathbb{R}^d)} := \|V_{g_0}(f)\|_{\boldsymbol{L}^1(\mathbb{R}^{2d})} \,. \tag{1.2}$$

We do not list all the good properties of this Banach algebra here, because this has been done meanwhile in many books and papers, but rather make the attempt to provide a guided tour through the meanwhile extensive literature on the subject, with some hints what has been already established, and what still has to be done.

This Segal algebra (in Reiter's sense) has been introduced (in a different fashion) in [10], and widely used by the author and many colleagues, especially those working in Time-Frequency Analysis (**TFA**).

Note that $(S_0'(\mathbb{R}^d), \|\cdot\|_{S_0'})$ could be introduced as the subspace of the space $\boldsymbol{S}'(\mathbb{R}^d)$ of tempered distributions with a bounded *spectrogram*, and with the norm

$$\|\sigma\|_{S_0'(\mathbb{R}^d)} := \sup_{(t,s)\in\mathbb{R}^d\times\widehat{\mathbb{R}}^d} |V_{g_0}(\sigma)(t,s)|.$$

In addition to the norm the w^*-convergence in $S_0'(\mathbb{R}^d)$ is highly important. It can be characterized as pointwise or in fact uniform convergence of $V_{g_0}(\sigma_n)(t,s)$ over compact subsets of *phase space* $\mathbb{R}^d \times \widehat{\mathbb{R}}^d$. At the level of *Gabor expansions* it corresponds to coordinate-wise convergence of the canonical Gabor coefficients (which in turn implies distributional convergence).

For a list of basic properties of S_0 (in the LCA context) see [44]. Many applications of these spaces have been given already in [40] and [35], published in the book [37] on Gabor Analysis (see also [38], with a contribution to Gabor multipliers provided by [36]).

A good way to understand the properties of this BGTr is to note that one can establish natural isomorphisms between the triple $(S_0, \boldsymbol{L}^2, S_0')(\mathbb{R}^d)$ and $(\boldsymbol{\ell}^1, \boldsymbol{\ell}^2, \boldsymbol{\ell}^\infty)(\mathbb{Z}^d \times \mathbb{N}^d)$ via the so-called *Wilson bases* which have been introduced in

[7], see [41]. This is rather analogous to the situation arising in the classical Fourier series expansions for the torus, which established not only a unitary operator from $\big(\boldsymbol{L}^2(\mathbb{T}), \|\cdot\|_2\big)$ to $\boldsymbol{\ell}^2(\mathbf{Z}_N)$. The restriction to Wiener's algebra $\big(\boldsymbol{A}(\mathbb{T}), \|\cdot\|_{\boldsymbol{A}}\big)$ of absolutely convergent Fourier series together with the extension (e.g. by duality) to the space $\boldsymbol{PM}(\mathbb{T})$ of pseudo-measures allows to consider the classical Fourier transform as a *unitary BGTr isomorphism* between $(\boldsymbol{A}, \boldsymbol{L}^2, \boldsymbol{PM})(\mathbb{T})$ and $(\boldsymbol{\ell}^1, \boldsymbol{\ell}^2, \boldsymbol{\ell}^\infty)(\mathbb{Z})$.

For the Euclidean setting we can consider the Fourier transform, defined on $S_0(\mathbb{R}^d)$ via Riemann integrals and extended to $SOPRd$ via

$$\widehat{\sigma}(f) = \sigma(\widehat{f}), \quad f \in S_0(\mathbb{R}^d).$$

The Fourier transform is a *unitary BGTr automorphism* of $(S_0, \boldsymbol{L}^2, S_0')(\mathbb{R}^d)$, which is characterized by the fact that it maps pure frequencies $\chi_s(t) = exp(2\pi ist) \in S_0'(\mathbb{R}^d)$ into corresponding Dirac measures $\delta_s \in S_0'(\mathbb{R}^d)$, while being unitary at the Hilbert space level. The fact that it is an adjoint operator also guarantees that it preserves w^*–convergence of bounded nets (which grants the uniqueness).

From our viewpoint the triple $(S_0, \boldsymbol{L}^2, S_0')(\mathbb{R}^d)$, the so-called *Banach Gelfand Triple* (**BGTr**) can be considered to be a key player for many considerations in Fourier and TF-Analysis. Although arising as an example if Segal algebra with specific (minimality) properties it turned out to provide the correct reservoir for Gabor analysis. Explaining this aspect in detail would require a separate survey article. But it is also relevant for the classical Fourier transform, in the sense that it is a valid domain for the Poisson Summation formula, valid for arbitrary discrete and co-compact subgroups and any $f \in S_0(\mathbb{R}^d)$, but also because according to [39] all the classical summability kernels belong to $S_0(\mathbb{R}^d)$. In fact, section 6 of that paper provides a list of 17 families of summability kernels which are all found to belong to members of $S_0(\mathbb{R}^d)$. Thus multiplying the Fourier transform of a function in $\boldsymbol{L}^1(\mathbb{R}^d)$ by such a summability kernel g we can be sure that $\widehat{f} \cdot g \in \mathcal{F}\boldsymbol{L}^1(\mathbb{R}^d) \cdot S_0(\mathbb{R}^d) \subseteq S_0(\mathbb{R}^d)$ and thus the inverse Fourier transform can be applied. By showing that $\mathcal{F}^{-1}(g)$ forms an approximate unit in $\big(\boldsymbol{L}^1(\mathbb{R}^d), \|\cdot\|_1\big)$ it is clear that one can recover in this way f from $\widehat{f}$.

2 Fourier Standard Spaces

The Banach Gelfand Triple also opens the way to what has been called by the author *Fourier Standard Spaces* (**FouSSps**). This family of Banach spaces of mild distributions comprises a large variety of Banach spaces arising naturally in Fourier and TF-Analysis.

Classical Fourier Analysis puts a lot of emphasis on $\boldsymbol{L}^p$-spaces for $1 \leq p \leq \infty$ or their natural generalizations, such as Lorentz or Orlicz spaces, which are mostly *rearrangement invariant* Banach spaces of (equivalence classes) of *measurable*, or better *locally integrable* functions. Y. Katznelson [46] emphasizes the role of *homogeneous Banach spaces*, which are Banach spaces of locally integrable functions on which the *regular representation* of the group by translates acts strongly continuous and isometrically (see also [54]). Homogeneous Banach spaces are in a natural way Banach modules over $(\boldsymbol{M}_b(G), \|\cdot\|_{\boldsymbol{M}_b})$ (with respect to convolution), see [27]. A special subclass of such homogeneous Banach spaces are the so-called *Segal algebras* introduced by H. Reiter (see [51, 52] or [53]). The are exactly those homogeneous Banach spaces which are embedded into $\left(\boldsymbol{L}^1(G), \|\cdot\|_1\right)$ as dense subspaces. As a consequence of the general principle (refreshed in [27], based on an elementary redefinition of convolution of bounded measures as given in [22]). In particular, any Segal algebra is an (essential) *Banach ideal* in $\left(\boldsymbol{L}^1(\mathbb{R}^d), \|\cdot\|_1\right)$ and thus a Banach algebra with respect to convolution.

Returning to the traditional Lebesgue spaces $\left(\boldsymbol{L}^p(\mathbb{R}^d), \|\cdot\|_p\right)$, with $1 \leq p < \infty$, which are homogeneous Banach spaces (for $p = \infty$ one should use $\left(\boldsymbol{C}_0(\mathbb{R}^d), \|\cdot\|_\infty\right)$) let us observe that unfortunately they show a different behaviour locally and globally and thus there are no inclusion relations between them. While on their subspaces of regular step functions (or in a similar way cubic spline functions on $\mathbb{R}$) the inclusions go in one direction (according to the behaviour of the spaces $(\boldsymbol{\ell}^p, \|\cdot\|_p)$) they go in the opposite direction over any compact subset of $\mathbb{R}^d$. This is the main motivation to study the *Wiener amalgam spaces* $\left(\boldsymbol{W}(\boldsymbol{L}^p, \boldsymbol{\ell}^q)(\mathbb{R}^d), \|\cdot\|_{\boldsymbol{W}(\boldsymbol{L}^p,\boldsymbol{\ell}^q)(\mathbb{R}^d)}\right)$, with *local component* $\boldsymbol{L}^p$ and *global component* $\left(\boldsymbol{\ell}^q, \|\cdot\|_{\boldsymbol{\ell}^q}\right)$. In addition to the little known Hausdorff-Young Theorem for amalgams (as provided in [9] one should recall the convolution, pointwise multiplier and interpolation results which are valid, even in the much more general context that has been introduced already in 1980 in the paper [11], but also [31] and [13]. The paper [11] actually extends the classical concept by allowing much more general local and global components $(\boldsymbol{B}, \|\cdot\|_{\boldsymbol{B}})$ and $(\boldsymbol{C}, \|\cdot\|_{\boldsymbol{C}})$ respectively, allowing to form the *Wiener amalgam spaces* $\left(\boldsymbol{W}(\boldsymbol{B}, \boldsymbol{C}), \|\cdot\|_{\boldsymbol{W}(\boldsymbol{B},\boldsymbol{C})}\right)$. This paper provides a description of the equivalence of continuous description (using so-called *control functions*, describing the local behaviour of a function or distribution) and the corresponding discrete characterization using the so-called **BUPUs** $\Psi = (\psi_i)_{i\in I}$, the *bounded uniform partitions of unity*, which play also crucial role in the rebuilding of Fourier Analysis from scratch. Here boundedness refers to the boundedness in the pointwise multiplier algebra of the local component $(\boldsymbol{B}, \|\cdot\|_{\boldsymbol{B}})$, and uniformity refers to the uniform size of the support of the functions ψ_i. A very recent summary of the "ubiquitous" appearance of BUPUs is given in the survey article [28].

Now FouSSPs are essentially Banach spaces of tempered distributions, sandwiched between $\boldsymbol{\mathcal{S}}(\mathbb{R}^d)$ and $\boldsymbol{\mathcal{S}}'(\mathbb{R}^d)$ (tempered distributions), i.e. satisfying

$$\boldsymbol{\mathcal{S}}(\mathbb{R}^d) \hookrightarrow (\boldsymbol{B}, \|\cdot\|_{\boldsymbol{B}}) \hookrightarrow \boldsymbol{\mathcal{S}}'(\mathbb{R}^d)$$

with the extra property that they are double modules with respect to $\left(\boldsymbol{L}^1(\mathbb{R}^d), \|\cdot\|_1\right)$ (under convolution) as well as over the Fourier Algebra $\left(\mathcal{F}\boldsymbol{L}^1(\mathbb{R}^d), \|\cdot\|_{\mathcal{F}\boldsymbol{L}^1}\right)$ with respect to pointwise multiplication. As a matter of fact $\left(S_0(\mathbb{R}^d), \|\cdot\|_{S_0}\right)$ is the smallest member in this family (this was the main motivation for its introduction in [10]), and any of these spaces is in fact continuously embedded into $S_0'(\mathbb{R}^d)$. In other words, we have automatically (and without loss of generality) a sandwich situation of the form

$$\left(S_0(\mathbb{R}^d), \|\cdot\|_{S_0}\right) \hookrightarrow (\boldsymbol{B}, \|\cdot\|_{\boldsymbol{B}}) \hookrightarrow (S_0'(\mathbb{R}^d), \|\cdot\|_{S_0'}).$$

Sometimes $S_0(\mathbb{R}^d)$ is dense (if and only if both module actions are *essential*), but the w^*–density of $S_0(\mathbb{R}^d)$ in $S_0'(\mathbb{R}^d)$ implies that $\boldsymbol{B}$ is always w^*–dense in $S_0'(\mathbb{R}^d)$. In fact, the *double module structure* of these spaces allows to generate a whole *diagram* of FouSSps which all have the same norm. If $(\boldsymbol{B}, \|\cdot\|_{\boldsymbol{B}})$ is reflexive as a Banach space the diagram collapses, but for $(\boldsymbol{B}, \|\cdot\|_{\boldsymbol{B}}) = \left(C_0, \|\cdot\|_\infty\right)$ (or equivalently) $\left(\boldsymbol{L}^\infty(\mathbb{R}^d), \|\cdot\|_\infty\right)$ the diagram shows 6 different spaces, all closed in $\left(\boldsymbol{L}^\infty(\mathbb{R}^d), \|\cdot\|_\infty\right)$ (see [4] for the details). The study of such a diagram for more sophisticated members of the family of FouSSps is an ongoing tasks with many open questions.

Note that among others the Fourier image of a FouSSp is a FouSSp as well, the spaces $\left(\mathcal{F}\boldsymbol{L}^p(\mathbb{R}^d), \|\cdot\|_{\mathcal{F}\boldsymbol{L}^p}\right)$ is in this family for any $p \in [1, \infty]$. Also Wiener amalgam spaces of the form $\left(\boldsymbol{W}(\boldsymbol{B}, \boldsymbol{C}), \|\cdot\|_{\boldsymbol{W}(\boldsymbol{B},\boldsymbol{C})}\right)$, with $(\boldsymbol{B}, \|\cdot\|_{\boldsymbol{B}})$ just any FouSSp and $(\boldsymbol{C}, \|\cdot\|_{\boldsymbol{C}}) = \left(\boldsymbol{L}^q(\mathbb{R}^d), \|\cdot\|_q\right)$ is another such spaces. Taking the inverse Fourier transform of a space such as $\boldsymbol{W}(\mathcal{F}\boldsymbol{L}^p, \boldsymbol{\ell}^q)(\mathbb{R}^d)$ gives exactly the *modulation spaces* $\left(\boldsymbol{M}^{p,q}(\mathbb{R}^d), \|\cdot\|_{\boldsymbol{M}^{p,q}}\right)$, as introduced in [16] (going back to the technical report of 1983, [12]).

Different variants of modulation spaces (see [17]) are nowadays widely used in different areas, including the theory of pseudo-differential operators. Two recent books in this direction are [2] and [5].

The so-called *Herz algebras* $\left(\boldsymbol{A}_p(\mathbb{R}^d), \|\cdot\|_{\boldsymbol{A}_p(\mathbb{R}^d)}\right)$ arise by taking a so-called *projective convolution tensor product* of $\left(\boldsymbol{L}^p(\mathbb{R}^d), \|\cdot\|_p\right)$ with its dual space $\left(\boldsymbol{L}^q(\mathbb{R}^d), \|\cdot\|_q\right)$. This space is another FouSSP and so it dual, which can be naturally identified with the space of translation invariant operators on $\left(\boldsymbol{L}^p(\mathbb{R}^d), \|\cdot\|_p\right)$. It has been pointed out in [29] that this can be seen as a consequence of two relatively simple observations: The embedding of Banach spaces $\left(\boldsymbol{A}_p(\mathbb{R}^d), \|\cdot\|_{\boldsymbol{A}_p(\mathbb{R}^d)}\right) \hookrightarrow \left(C_0(\mathbb{R}^d), \|\cdot\|_\infty\right)$ allows to identify a translation invariant operator T on $\left(\boldsymbol{A}_p(\mathbb{R}^d), \|\cdot\|_{\boldsymbol{A}_p(\mathbb{R}^d)}\right)$ by some functional $\sigma \in \boldsymbol{PM}_p(\mathbb{R}^d) = \boldsymbol{A}_p(\mathbb{R}^d)'$, via $\sigma(f) = T(f^\checkmark)(0)$, where $f^\checkmark(x) = f(-x)$ is the usual reflection operator, and secondly, by verifying that there is a natural one-to-one relationship between these "multipliers" of $\left(\boldsymbol{A}_p(\mathbb{R}^d), \|\cdot\|_{\boldsymbol{A}_p(\mathbb{R}^d)}\right)$ and those of $\left(\boldsymbol{L}^p(\mathbb{R}^d), \|\cdot\|_p\right)$ (see also [8] for a very detailed investigation of the situation).

3 The Frame-Work of Conceptual Harmonic Analysis

Thoughts on Conceptual Harmonic Analysis have been already formulated in [21]. The idea is to combine the benefits of Abstract Harmonic Analysis (**AHA**) with those of *Numerical* or *Computational Harmonic Analysis.* This asks for a combination of methods from the theory of *function spaces* (see [19]) and *approximation theory* with computational methods. Early version of this idea are given in [24] or [20]. Typically the function spaces view-point allows to derive robustness of the ingredients, e.g. to approximate problems in Gabor analysis based on general lattices to those with rational lattice constants (using $\left(S_0(\mathbb{R}^d), \|\cdot\|_{S_0}\right)$ windows), as described in detail in [34], or [15]. On the other hand the group theoretic point of view allows to relate questions in Gabor analysis to similar questions already better understood (see [14]), but this group-theoretical approach also opens the way to do the analogue in a computational (linear algebra) context, as a first stop towards a *structure preserving approximation.* In fact, the S_0-setting allows to closely relate (again, using BUPUs) continuous problems to corresponding discrete ones, by periodization and sampling, while BUPUs allow to approximately recover from (sufficiently many fine) samples a function in $S_0(\mathbb{R}^d)$ (see [34, 45]).

4 Banach Gelfand Triples

A list of references concerning the specific Banach Gelfand Triples (BGTr) reads as follows: While implicitly that the foundations have been laid down in [40] and [35] the formal description was substantiated in [6], or the master thesis [1]. For application aspects see [18] or the survey emphasizing various applications given in [23]. The BGTr also plays an important role in the design of new course material for both engineers and mathematicians, as outlined in [26], and has been tested in courses at ETH Zurich (2015 and 2020) and TU Munich (2017 and 2018). The material related to the last ETH course can be retrieved from www.nuhag.eu/ETH20.

A summary of the aspects of the BGTr in the context of sampling theory has been given in the contribution to the R. Higgins Memorial Book [30]. Two papers showing that one can introduce the BGTr using elementary are given by [32], and in the same year [25] featuring a *sequential approach,* inspired by the Lighthill approach [47] which was supposed to make the Schwartz theory of tempered distributions accessible to engineers. In fact, it is very easy, using the properties of regularizing operators, to show via the Banach Steinhaus principle that $(S_0'(\mathbb{R}^d), \|\cdot\|_{S_0'})$ can be identified with the space of all **ECMiCS**, i.e. *Equivalence Classes of Mild Cauchy Sequences.* The general study of such regularization operators, which is of linear operators with kernels in $S_0(\mathbb{R}^{2d})$ (or equivalently mapping bounded, w^*–convergent sequences in $(S_0'(\mathbb{R}^d), \|\cdot\|_{S_0'})$ into norm convergent sequences in $\left(S_0(\mathbb{R}^d), \|\cdot\|_{S_0}\right)$) is given in [33]. It is related to the so-called *inner kernel theorem.* In fact, it is part of the so-called *BGTr kernel Theorem* (more or less given already in [35]) which described a natural BGTr isomorphism between the (integral) kernels in $(S_0, \boldsymbol{L}^2, S_0')(\mathbb{R}^{2d})$ and the corresponding operator BGTr $\big(\mathcal{L}(S_0', S_0), \mathcal{HS}, \mathcal{L}(S_0, S_0')\big)$, describing the corresponding Banach spaces of linear operators, with the Hilbert space of Hilbert Schmidt operators (on $\big(\boldsymbol{L}^2(\mathbb{R}^d), \|\cdot\|_2\big)$) in the middle, endowed with the usual scalar product $\langle T, S\rangle_{\mathcal{HS}} := \mathrm{trace}(TS^*)$. Note that for applications to Fourier or TF-Analysis this BGTr of operators appears to be more suitable than the usual one $(\mathcal{T}^1, \mathcal{HS}, \mathcal{L}(\boldsymbol{L}^2))(\mathbb{R}^d)$ of trace class, Hilbert Schmidt and bounded operators on $\mathcal{H} = \big(\boldsymbol{L}^2(\mathbb{R}^d), \|\cdot\|_2\big)$. Among others $\mathcal{L}(S_0, S_0')$ contains any bounded operator between pairs of arbitrary spaces $\big(\boldsymbol{L}^p(\mathbb{R}^d), \|\cdot\|_p\big)$, with $1 \le p < \infty$, including those which do not need to be bounded on $\big(\boldsymbol{L}^2(\mathbb{R}^d), \|\cdot\|_2\big)$.

Most recently a number of investigations in the field of *Quantum Harmonic Analysis* (as introduced by R. Werner in [55]) make use of the tools mentioned in this survey, and there is more to come (see e.g. [3, 43, 48–50]) or [42].

5 Final Remarks

Almost all the papers by the author can be retrieved from https://nuhagphp.univie.ac.at/home/feipub_db.php.

Acknowledgments This note has been written during a short stay at CIRM in Luminy in November 2023, celebrating the 10th anniversary of the Morlet Chair. The author held this position in the autumn of 2014.

References

1. S. Bannert, Banach Gelfand Triples and applications in time-frequency analysis. Master's thesis, University of Vienna, 2010
2. A. Benyi, K.A. Okoudjou, *Modulation Spaces with Applications to Pseudodifferential Operators and Nonlinear Schrodinger Equations*. Appl. Num. Harm. Anal. (ANHA) Series (Springer (Birkhauser), New York, 2020)
3. E. Berge, S. Berge, and R. Fulsche. A quantum harmonic analysis approach to Segal algebras. Preprint. arXiv:2301.09384 (2023)
4. W. Braun, H.G. Feichtinger, Banach spaces of distributions having two module structures. J. Funct. Anal. **51**, 174–212 (1983)
5. E. Cordero, L. Rodino, *Time-frequency Analysis of Operators and Applications*, De Gruyter Studies in Mathematics (De Gruyter, Berlin, 2020)
6. E. Cordero, H.G. Feichtinger, F. Luef, Banach Gelfand triples for Gabor analysis, in *Pseudo-differential Operators*, volume 1949 of Lect. Notes Math. (Springer, Berlin, 2008), pp. 1–33
7. I. Daubechies, S. Jaffard, J.L. Journé, A simple Wilson orthonormal basis with exponential decay. SIAM J. Math. Anal. **22**, 554–573 (1991)
8. A. Derighetti, *Convolution Operators on Groups* (Springer Berlin/Heidelberg, 2011)
9. H.G. Feichtinger, Banach spaces of distributions of Wiener's type and interpolation, in *Proc. Conf. Functional Analysis and Approximation, Oberwolfach August 1980*, ed. by P. Butzer, S. Nagy, E. Gorlich, number 69, Internat. Ser. Numer. Math. (Birkhauser Boston, Basel, 1981), pp. 153–165

10. H.G. Feichtinger, On a new Segal algebra. Monatsh. Math. **92**, 269–289 (1981)
11. H.G. Feichtinger, Banach convolution algebras of Wiener type, in *Proc. Conf. on Functions, Series, Operators, Budapest 1980*, ed. by B. Sz.-Nagy, J. Szabados, volume 35 of Colloq. Math. Soc. Janos Bolyai (North-Holland, Amsterdam, 1983), pp. 509–524
12. H.G. Feichtinger, Modulation spaces on locally compact Abelian groups. Technical report, University of Vienna, January 1983
13. H.G. Feichtinger, Banach spaces of distributions defined by decomposition methods. II. Math. Nachr. **132**, 207–237 (1987)
14. H.G. Feichtinger, Spline-type spaces in Gabor analysis, in *Wavelet Analysis: Twenty Years Developments Proceedings of the International Conference of Computational Harmonic Analysis, Hong Kong, China, June 4–8, 2001*, ed. by D.X. Zhou, volume 1 of Ser. Anal. (World Scientific Publishing, River Edge, 2002), pp. 100–122
15. H.G. Feichtinger, Gabor multipliers with varying lattices, in *Proc. SPIE Conf.*, August (2003), p. 14
16. H.G. Feichtinger, Modulation spaces on locally compact Abelian groups, in *Proc. Internat. Conf. on Wavelets and Applications*, ed. by R. Radha, M. Krishna, S. Thangavelu, Chennai, January 2002 (Allied Publishers, New Delhi, 2003), pp. 1–56
17. H.G. Feichtinger, Modulation spaces: looking back and ahead. Sampl. Theory Signal Image Process. **5**(2), 109–140 (2006)
18. H.G. Feichtinger, Banach Gelfand triples for applications in physics and engineering, in volume 1146 of *AIP Conf. Proc.* (American Institute of Physics, 2009), pp. 189–228
19. H.G. Feichtinger, *Choosing Function Spaces in Harmonic Analysis*, volume 4 of The February Fourier Talks at the Norbert Wiener Center, Appl. Numer. Harmon. Anal. (Birkhauser/Springer, Cham, 2015), pp. 65–101
20. H.G. Feichtinger, Elements of postmodern harmonic analysis, in *Operator-related Function Theory and Time-Frequency Analysis. The Abel Symposium 2012, Oslo, Norway, August 20–24, 2012* (Springer, Cham, 2015), pp. 77–105
21. H.G. Feichtinger, Thoughts on numerical and conceptual harmonic analysis, in *New Trends in Applied Harmonic Analysis. Sparse Representations, Compressed Sensing, and Multifractal Analysis*, ed. by A. Aldroubi, C. Cabrelli, S. Jaffard, U. Molter. Applied and Numerical Harmonic Analysis (Birkhauser, Cham, 2016), pp. 301–329
22. H.G. Feichtinger, A novel mathematical approach to the theory of translation invariant linear systems, in *Recent Applications of Harmonic Analysis to Function Spaces, Differential Equations, and Data Science.*, ed. by I. Pesenson, Q. Le Gia, A. Mayeli, H. Mhaskar, D. Zhou. Applied and Numerical Harmonic Analysis (Birkhauser, Cham, 2017), pp. 483–516
23. H.G. Feichtinger, Banach Gelfand triples and some applications in harmonic analysis, in *Proc. Conf. Harmonic Analysis (Abidjan, May 2018)*, ed. by J. Feuto, M. Essoh (2018), pp. 1–21
24. H.G. Feichtinger, Classical Fourier Analysis via mild distributions. MESA Non-linear Stud. **26**(4), 783–804 (2019)
25. H.G. Feichtinger, A sequential approach to mild distributions. Axioms **9**(1), 1–25 (2020)
26. H.G. Feichtinger, Ingredients for applied fourier analysis, in *Sharda Conference Feb. 2018* (Taylor and Francis, 2020), pp. 1–22
27. H.G. Feichtinger, Homogeneous Banach spaces as Banach convolution modules over $M(G)$. Mathematics **10**(3), 1–22 (2022)
28. H. Feichtinger, The ubiquitous appearance of BUPUs, in *Proc. Conference, Abidjan 2023*, ed. by J. Feuto et al., Appl. Num. Harm. Analysis (Birkhauser/Springer, Cham, 2023)
29. H.G. Feichtinger, A characterization of multipliers of the Herz Algebra. Axioms **12**, 482 (2023)
30. H.G. Feichtinger, Sampling via the Banach Gelfand triple, in *Sampling, Approximation, and Signal Analysis Harmonic Analysis in the Spirit of J. Rowland Higgins*, ed. by S.D. Casey, M. Dodson, P. J.S.G. Ferreira, A. Zayed. Appl. Num. Harm. Anal. (Taylor & Francis, 2023), pp. 1–30
31. H.G. Feichtinger, P. Grobner, Banach spaces of distributions defined by decomposition methods. I. Math. Nachr. **123**, 97–120 (1985)

32. H.G. Feichtinger, M.S. Jakobsen, Distribution theory by Riemann integrals, in *Mathematical Modelling, Optimization, Analytic and Numerical Solutions* (2020), pp. 33–76
33. H.G. Feichtinger, M.S. Jakobsen, The inner kernel theorem for a certain Segal algebra. Monatsh. Math. **198**(4), 675–715 (2022)
34. H.G. Feichtinger, N. Kaiblinger, Quasi-interpolation in the Fourier algebra. J. Approx. Theory **144**(1), 103–118 (2007)
35. H.G. Feichtinger, W. Kozek, Quantization of TF lattice-invariant operators on elementary LCA groups, in *Gabor Analysis and Algorithms*, ed. by H.G. Feichtinger, T. Strohmer, Appl. Numer. Harmon. Anal. (Birkhauser, Boston, 1998), pp. 233–266
36. H.G. Feichtinger, K. Nowak, A first survey of Gabor multipliers, in *Advances in Gabor Analysis*, ed. by H.G. Feichtinger, T. Strohmer, Appl. Numer. Harmon. Anal. (Birkhauser, 2003), pp. 99–128
37. H.G. Feichtinger, T. Strohmer, *Gabor Analysis and Algorithms. Theory and Applications* (Birkhauser, Boston, 1998)
38. H.G. Feichtinger, T. Strohmer, *Advances in Gabor Analysis* (Birkhauser, Basel, 2003)
39. H.G. Feichtinger, F. Weisz, The Segal algebra $S_0(R^d)$ and norm summability of Fourier series and Fourier transforms. Monatsh. Math. **148**, 333–349 (2006)
40. H.G. Feichtinger, G. Zimmermann, A Banach space of test functions for Gabor analysis, in *Gabor Analysis and Algorithms: Theory and Applications*, ed. by H.G. Feichtinger, T. Strohmer. Applied and Numerical Harmonic Analysis (Birkhauser Boston, 1998), pp. 123–170
41. H.G. Feichtinger, K. Grochenig, D.F. Walnut, Wilson bases and modulation spaces. Math. Nachr. **155**, 7–17 (1992)
42. H. Feichtinger, S. Halvdansson, F. Luef, Measure-operator convolutions and applications to mixed-state Gabor multipliers. Preprint. arXiv:2308.04985 (2023)
43. R. Fulsche, N. Galke, Quantum Harmonic Analysis on locally compact abelian groups. Preprint. arXiv:2308.02078 (2023)
44. M.S. Jakobsen, On a (no longer) New Segal Algebra: a review of the Feichtinger algebra. J. Fourier Anal. Appl. **24**(6), 1579–1660 (2018)
45. N. Kaiblinger, Approximation of the Fourier transform and the dual Gabor window. J. Fourier Anal. Appl. **11**(1), 25–42 (2005)
46. Y. Katznelson, *An Introduction to Harmonic Analysis*, 2nd corr. edn. (Dover Publications, New York, 1976)
47. M.J. Lighthill, *Introduction to Fourier Analysis and Generalised Functions*, Students' Edition (Cambridge University Press, Cambridge, 1962)
48. F. Luef, E. Skrettingland, Convolutions for Berezin quantization and Berezin-Lieb inequalities. J. Math. Phys. **59**(2), 023502, 11 (2018)
49. F. Luef, E. Skrettingland, Convolutions for localization operators. J. Math. Pures Appl. (9) **118**, 288–316 (2018)
50. F. Luef, E. Skrettingland, A Wiener Tauberian theorem for operators and functions. J. Funct. Anal. **280**(6), 108883 (2021)
51. H. Reiter, *Classical Harmonic Analysis and Locally Compact Groups* (Clarendon Press, Oxford, 1968)
52. H. Reiter, *L^1-algebras and Segal Algebras* (Springer, Berlin, 1971)
53. H. Reiter, J.D. Stegeman, *Classical Harmonic Analysis and Locally Compact Groups*, 2nd edn. (Clarendon Press, Oxford, 2000)
54. H.S. Shapiro, *Topics in Approximation Theory*, volume 187 of Lecture Notes in Mathematics (Springer, Berlin (1971)
55. H.J. Weaver, *Applications of Discrete and Continuous Fourier Analysis* (Wiley-Interscience, 1983)

A Note on a Frictional Unilateral Contact Problem in Nonlinear Elasticity

Joachim Gwinner

Abstract This note is devoted to a frictional unilateral contact problem in finite strain elasticity, where the Tresca friction model is adapted from linear elasticity. Here we focus to the issue of existence and aim at closing a gap in Gwinner (Philos Trans R Soc A 380(2236):Paper No. 20210356, 2022) between modeling and the existence proof using recession analysis together with the polyconvexity approach to nonlinear elasticity due to J. Ball (Arch Rational Mech Anal 63:337–403, 1977). As in Gwinner (Philos Trans R Soc A 380(2236):Paper No. 20210356, 2022) we include the delicate case where the elastic body is neither fixed nor a deformation is prescribed along some part of its boundary, but the body rests on a rigid foundation with a free boundary and is only submitted to forces and loads, acting in the interior (like gravity) and at the boundary, respectively. This leads to a loss of coercivity and necessitates an extra condition such that the geometry of the obstacle prevents the body to escape. This new condition extends a similar condition of Ciarlet and Nečas (Arch Rational Mech Anal 87:319–338, 1985) from the frictionless case to the case of Tresca friction.

Keywords Tresca friction · Unilateral contact · Finite strain

2000 Mathematics Subject Classification 74B20,74G22,74M10,74M15

1 Introduction

This note is devoted to existence theory for a frictional unilateral contact problem in finite strain elasticity, where the Tresca friction model, which is a simplified version of the more involved Coulomb friction model, as explained later in the text, is

---

J. Gwinner (✉)
Institute of Applied Mathematics, Department of Aerospace Engineering,
Universität der Bundeswehr München, Neubiberg, Germany
e-mail: joachim.gwinner@unibw-muenchen.de

R. Duduchava et al. (eds.), *Tbilisi Analysis and PDE Seminar*,
Research Perspectives Ghent Analysis and PDE Center 7,
https://doi.org/10.1007/978-3-031-62894-8_8

adapted from linear elasticity. Here we aim at closing a gap in [8] between modeling and the proof of existence using recession analysis together with the polyconvexity approach to nonlinear elasticity due to J. Ball [2]. As in [8] the existence theory includes the delicate case where the elastic body is neither fixed nor any deformation is prescribed along some part of its boundary, but the body lies on a rigid foundation with a free boundary and is only submitted to forces and loads, acting in the interior (like gravity) and at the boundary, respectively. This leads to a loss of coercivity ("semicoercive problem") and necessitates an extra condition such that the geometry of the obstacle due to the foundation prevents the body to escape. This new condition extends a similar condition of Ciarlet and Nečas [4] from the frictionless case to the case of Tresca friction.

The plan of the note is as follows. The subsequent Sect. 2 formulates the frictional unilateral contact problem and provides some preliminary material for the existence theory. Then Sect. 3 presents the existence Theorem 3.1 modifying the proof of [8].

2 The Frictional Unlateral Contact Problem: The Setting and Some Preliminaries

Let Ω be a bounded open connected subset of $\mathbb{R}^d$ $(d = 2, 3)$ with a Lipschitz continuous boundary $\Gamma = \partial\Omega$ with the outward unit normal n to Γ defined almost everywhere (a.e.).

Following [3] for the framework of finite elasticity, the set $\overline{\Omega}$ is the *reference configuration* occupied by a deformable elastic body in the absence of any applied force. *Deformations* of this body are given by functions $\psi : \overline{\Omega} \to \mathbb{R}^d$ which belong to some Sobolev space $W^{1,p}(\Omega, \mathbb{R}^d)$ [1] for some $p \in (1, \infty)$.

Consider the gradient mapping $\nabla\psi : \Omega \mapsto M^d$ (the set of all matrices of order d). With adj A and det A denoting the adjugate (= transpose of the cofactor matrix) and the determinant of $A \in M^d$, respectively, introduce the set

$$X = \{(\psi, A, \delta) | A = \operatorname{adj} \nabla\psi, \delta = \det \nabla\psi\}$$

contained in

$$\mathbb{X} = W^{1,p}(\Omega, \mathbb{R}^d) \times L^q(\Omega, \mathbb{R}^{d\times d}) \times L^r(\Omega)$$

$(p, q, r$ specified later) and in addition the subset

$$V_> = \{\psi | (\psi, A, \delta) \in X, \det \nabla\psi > 0 \quad \text{a.e. in } \Omega\} ,$$

since we are only interested in at least locally invertible deformations. For incompressible material we further specify

$$V_= = \{\psi | (\psi, A, \delta) \in X, \det \nabla\psi = 1 \quad \text{a.e. in } \Omega\} .$$

This further constraint det $=$ 1 does not pose an extra mathematical difficulty. Therefore we can simply set $V := V_>$ for compressible material, respectively $V := V_=$ for incompressible material in what follows.

Observe that X is *not convex*, however it is a (sequentially) *weakly closed* subset of $\mathbb{X}$. This is the basic fact for existence and convergence results in nonlinear elasticity and is an immediate consequence of a result, due to J. Ball [2, Theorem 6.2]. With weak or weak* convergence of a sequence denoted by $\rightharpoonup$, this result reads as follows.

Proposition 2.17 ([2, Theorem 3.4, p 218]) *Let* $p \geq 2$. *If* $\psi^{(r)} \rightharpoonup \psi$ *in* $(W^{1,p}(\Omega), \mathbb{R}^d)$, *then* adj $\nabla\psi^{(r)} \rightharpoonup$ adj $\nabla\psi$ *in the dual space of distributions. If* $\psi^{(r)} \rightharpoonup \psi$ *in* $(W^{1,p}(\Omega), \mathbb{R}^d)$ *and* $\operatorname{adj}\nabla\psi^{(r)} \rightharpoonup \operatorname{adj}\nabla\psi$ *in* $(L^{p'}(\Omega), \mathbb{R}^{d\times d})$, $\frac{1}{p}+\frac{1}{p'} =$ 1), *then* $\det \nabla\psi^{(r)} \rightharpoonup \det \nabla\psi$ *in the dual space of distributions.*

Let us return to the elastic body loaded by a volume force density $f_0 \in L^{p'}(\Omega, \mathbb{R}^d)$. Further Γ decomposes as $\Gamma = \overline{\Gamma}_0 \cup \overline{\Gamma}_\tau \cup \overline{\Gamma}_c$, where Γ_0 is the Dirichlet boundary part—which may be empty, see later—with prescribed deformations $\psi_0 \in L^p(\Gamma, \mathbb{R}^d)$. On the Neumann boundary part Γ_τ we have the traction condition

$$\tilde{T}(\nabla\psi)n = \tau_0 \quad \text{on } \Gamma_\tau$$

with the given surface load density $\tau_0 \in L^{p'}(\Gamma_\tau, \mathbb{R}^d)$ (with the dual exponent $p' = \frac{p}{p-1}$), where the mapping

$$\tilde{T} : A \in M^d \mapsto \tilde{T}(A) := \frac{\partial W}{\partial A}(A) \in M^d$$

denotes the response function associated with the *first Piola-Kirchhoff stress tensor* and $W(x, \cdot)$ denotes the *stored energy function.*

As Ciarlet and Nečas [4] we follow Ball's approach to nonlinear elasticity [2] and assume that $W(x, \cdot)$ is *polyconvex*: There exists a function $\tilde{W} : \Omega \times M^d \times M^d \times]0, +\infty[\to \mathbb{R}$ such that for all $x \in \Omega$, $\tilde{W}(x, ., ., .)$ is convex and for all $F \in M^d_+$, i.e. for all $F \in M^d$ with $\det F > 0$,

$$W(x, F) = \tilde{W}(x, F, \operatorname{adj} F, \det F)$$

holds. Secondly the following behaviour of $\tilde{W}$ is assumed as det $F \to 0+$:

$$(H_1)\quad \left.\begin{array}{ll} F_n \to F & \text{in } M^d \\ H_n \to H & \text{in } M^d \\ 0 < \delta_n \to 0 & \end{array}\right\} \text{ imply } \tilde{W}(x, F_n, H_n, \delta_n) \to +\infty \qquad (x \in \Omega)\,.$$

This means that infinite energy is required to compress a volume to a point. Thirdly $\tilde{W}$ is assumed to be *coercive*: There exists $a \in L^1(\Omega)$, moreover reals $b > 0$, $p \geq 2$, $q \geq \frac{p}{p-1}$, $r > 1$ such that

$$(H_2)\quad \tilde{W}(x, F, H, \delta) \geq a(x) + b(\|F\|^p + \|H\|^q + \delta^r) \qquad (x \in \Omega)$$

holds for all $F \in M^d$, $H \in M^d$, $\delta > 0$, where $\|F\|$ denotes the Euclidean norm of a matrix F.

On the remaining boundary part Γ_c prescribe frictional unilateral conditions as follows. For that decompose the displacement field $u := \psi - id$, respectively the boundary stress Tn, into its normal component $u_n n$, $u_n := n \cdot u$ and its tangential component $u_t := u - u_n n$; respectively its normal component $\tau_n n$, $\tau_n := n \cdot Tn$ and its tangential component $\tau_t := Tn - \tau_n n$. On Γ_c we impose the unilateral condition $u_n \leq 0$ a.e. describing the nonpenetration of the elastic body into a rigid foundation. This gives rise to the convex closed subset

$$K = \left\{\psi \in W^{1,p}(\Omega, \mathbb{R}^d) \,:\, u_n \leq 0 \text{ a.e. on } \Gamma_c\right\}.$$

Moreover in the tangential direction, we adopt the Tresca friction (given friction) model, similar to the case of linear elasticity [6, 9], and introduce for some given $g \in L^\infty_+(\Gamma_c)$ the convex function

$$G(\psi) = \int_{\Gamma_c} g|u_t|\,.$$

Let us add that the Tresca friction model is a simplification of the more realistic Coulomb friction model, see the following remark.

Remark 2.18 The above convex function G that enters the variational problem below results from the boundary condition that describes Tresca friction in tangential direction in strong form, see [6, 9]

$$(TFC)\quad \begin{cases} |\tau_t| \leq g\,, \\ \text{if } |\tau_t| < g \text{ then } u_t = 0\,, \\ \text{if } |\tau_t| = g \text{ then } \exists \kappa \geq 0 : u_t = -\kappa \tau_t\,. \end{cases}$$

This condition evidently simplifies the following boundary condition which is known as Coulomb friction condition [6, 9]:

$$(CFC)\begin{cases} |\tau_t| \le \nu_F |\tau_n| \,, \\ \text{if } |\tau_t| < \nu_F |\tau_n| \text{ then } u_t = 0 \,, \\ \text{if } |\tau_t| = \nu_F |\tau_n| \text{ then } \exists \kappa \ge 0 : u_t = -\kappa \tau_t \,, \end{cases}$$

where ν_F is a given coefficient of friction.

Thus the *set of admissible deformation fields* is described by the weakly closed set $A := K \cap V_0$, where $V_0 = \{\psi \in V \subset W^{1,p}(\Omega, \mathbb{R}^d) \,:\, \psi|_{\Gamma_0} = \psi_0\}$, and we are led to the *total energy*

$$I(\psi) = \int_\Omega W(x, \nabla\psi) dx + G(\psi) - F(\psi) \,,$$

where the first integral represents the *strain energy* of the body and the linear form

$$\begin{aligned} F(\psi) &= (f_0, \psi) + \langle \tau_0, \psi \rangle \\ &= \int_\Omega f_0 \cdot \psi \, dx + \int_{\Gamma_\tau} \tau_0 \cdot \psi \, da \,, \end{aligned}$$

is the external work, done by the volume load f_0 and surface load τ_0,

In these terms the contact problem under study reads:

$$(\mathbb{P}) \qquad \text{Find a minimizer } \psi \in A \text{ of } I \text{ on } A! \tag{2.1}$$

Similar to Ciarlet and Nečas [4] in nonlinear elasticity and similar to Duvaut and Lions [5, §4.4.2] in linear elasticity, we admit the case of a pure contact-traction problem with formally meas $\Gamma_0 = \emptyset$. Then we do not obtain a coercive variational problem although W itself is coercive. However we obtain a semicoercive problems, which we can treat by a decomposition technique in the following way.

Proposition 2.19 ([8, Proposition 2.3]) *Consider the semi-norm*

$$|\psi|_{1,p} = (\int_\Omega \|\nabla\psi\|^p \, dx)^{1/p}$$

and the L^p–norm

$$\|\psi\|_p = (\int_\Omega \|\psi\|^p \, dx)^{1/p} \,.$$

Then:

(i) *The norm* $\| \cdot \|_{1,p}$ *on* $W^{1,p}(\Omega, \mathbb{R}^d)$ *is equivalent to* $| \cdot |_{1,p} + \| \cdot \|_p$.
(ii) $R := \{\rho \in W^{1,p}(\Omega; \mathbb{R}^d) | \, |\rho|_{1,p} = 0\}$ *is of finite dimension.*
(iii) *There exists some* $c_0 > 0$ *such that for all* $\psi \in W^{1,p}(\Omega; \mathbb{R}^d)$,

$$\inf_{\rho \in R} \|\psi - \rho\|_p \leq c_0 \, |\psi|_{1,p} \, .$$

Consequently there is a uniquely defined projection $\pi : \psi \in W^{1,p}(\Omega, \mathbb{R}^d) \to \rho = \pi \psi \in R$*, any* $\psi \in W^{1,p}(\Omega, \mathbb{R}^d)$ *decomposes as follows:*

$$\psi = \pi \psi + (\psi - \pi \psi) =: \rho + \sigma \, , \quad \rho \in R \, ,$$

and there holds

$$\|\sigma\|_p \leq c_0 \, |\sigma|_{1,p} \, .$$

Mechanically R represents the rigid translations, whereas in linear elasticity the rigid rotations are included, see e.g. [5, III (3.39)].

Let us underline that the hypotheses above, in particular the polyconvexity are satisfied by realistic models of elastic materials, in particular by rubber-like materials. Here we can refer to the seminal work of Ogden [10] and for a more detailed discussion and more recent references to [8, section 2].

3 An Existence Result

To ensure that $A \neq \emptyset$ in the case meas $\Gamma_0 > 0$, we assume that the datum ψ_0 can be extended to $\psi_0 \in V \subset W^{1,p}(\Omega, \mathbb{R}^d)$ with $\psi_{0n} \leq 0$ a.e. on Γ_c.

We emphasize that we admit here the most delicate case of no prescribed deformations; that is formally, meas $\Gamma_0 = 0$. To handle this case, we employ recession analysis [11, $ 8], that is, we introduce the recession function of the convex, but not positively homogeneous Tresca friction energy G,

$$G_\infty(\psi) := \lim_{\tau \to \infty} \frac{1}{\tau} \, [G(\psi_0 + \tau \psi) - G(\psi_0)] = \int_{\Gamma_c} g |\psi_t| \, ,$$

where ψ_0 can be arbitrarily chosen in the nonvoid set V_{ad}. By the simple inequality $||x| - |y|| \leq |x - y|$ and by continuity of trace map, we can estimate

$$(*) \quad G(\rho + \sigma) - G_\infty(\rho) \geq -\|g\|_\infty (\|\gamma\| \|\sigma\|_{1,p} + \text{ meas } \Gamma_c) \, .$$

Then analogously to the "pressure condition" in linear elasticity (cf. [5, III (5.97)], [6, (3.2.2)]), which is going back to Fichera [7], in this situation we impose the condition

$$(H_3) \quad F(\rho) - G_\infty(\rho) < 0 \quad \forall \rho \in R \cap K \backslash \{0\}\,.$$

This condition corresponds to the hypothesis (4.18) of Ciarlet and Nečas in the instance of frictionless contact with a convex closed cone [4, pp. 333–335], which describes that "dangerous" directions where the body can "escape" should have an obtuse angle with the applied forces.

Now we can present our existence result for the variational problem $\mathbb{P}$.

Theorem 3.1 *Let $\tilde{W}$ satisfy (H_1) and (H_2). Moreover let $2 \le p < 3$, $q \ge \frac{p}{p-1}$, $r > 1$, $p \le s \le \frac{2p}{3-p}$. Suppose either* meas $\Gamma_0 > 0$ *or else there holds (H_3). Then there exists a solution $\hat{\psi}$ to the problem $\mathbb{P}$.*

Proof Follow the standard pattern of existence proofs in calculus of variations: Establish a priori bounds and construct a solution by an infimizing sequence.

Step 1: *A priori bounds for $I(\psi)$ in the space $E := W^{1,p}(\Omega; \mathbb{R}^d)$*
In the case meas $\Gamma_0 > 0$, $|\cdot|_{1,p}$ is equivalent to the norm $\|\cdot\|_E = \|\cdot\|_{1,p}$ on E. and thus a priori bounds immediately follow from (H_2).
Therefore the case meas $\Gamma_0 = 0$ remains. Since R is of finite dimension and G_∞ is positively homogenous, (H_3) implies that

$$-(f_0, \rho) - \langle \tau_0, \rho \rangle + G_\infty(\rho) \ge c_0 \|\rho\|_E \tag{3.1}$$

holds for all $\rho \in R \cap K$, where c_0 and in what follows $c_1, \ldots$ are generic positive constants.
For any $\psi \in E$, by $\nabla \rho = 0$, using the decomposition $\psi = \rho + \sigma$ introduced above,

$$\begin{aligned} I(\psi) &= \int_\Omega W(x, \nabla \psi) dx - (f_0, \psi) - \langle \tau_0, \psi \rangle + G(\psi) \\ &= \int_\Omega W(x, \nabla \sigma) dx - (f_0, \sigma) - (f_0, \rho) - \langle \tau_0, \sigma \rangle - \langle \tau_0, \rho \rangle \\ &\quad + G(\rho + \sigma)\,, \end{aligned}$$

further by the coerciveness of $\tilde{W}$, and by the continuity of $(f_0, \cdot)$ and $\langle \tau_0, \cdot \rangle$,

$$\begin{aligned} I(\psi) \ge &-\int_\Omega |a| dx + b \int_\Omega \{\|\nabla \sigma\|^p + \|\text{adj}\, \nabla \sigma\|^q + (\det \nabla \sigma)^r\}\, dx \qquad (3.2) \\ &- c_1(f_0, \tau_0)\|\sigma\|_E - (f_0, \rho) - \langle \tau_0, \rho \rangle + G(\sigma + \rho)\,, \end{aligned}$$

further by (3.1) and by (*),

$$I(\psi) \geq -c_a + b \int_\Omega \{\|\nabla\sigma\|^p + \|\text{adj } \nabla\sigma\|^q + (\det \nabla\sigma)^r\}\, dx$$
$$-c_1\|\sigma\|_{1,p} + c_0\|\rho\|_{1,p} - c_1(g, \gamma)\, \|\sigma\|_{1,p} - c_2(g)\,.$$

Therefore in virtue of Proposition 2.19 (i), for any $\psi \in V_{\text{ad}}$,

$$I(\psi) \geq -\tilde{c}_a + b|\sigma|_{1,p}^p + \tilde{c}_0\|\rho\|_p - \tilde{c}_1(|\sigma|_{1,p} + \|\sigma\|_p)\,. \tag{3.3}$$

Step 2: *Infimizing sequence* (ψ^k) *possesses weak limit points in the space* E
Consider an infimizing sequence $(\psi^k)_{k\in\mathbb{N}}$. By Proposition 2.19 (iii), $\|\sigma\|_p \leq c_2\|\psi\|_{1,p}$ and hence by $|\sigma|_{1,p} = |\psi|_{1,p}$, (3.3) shows in particular

$$I(\psi^1) \geq I(\psi^k) \geq -\tilde{c}_a + b|\psi^k|_{1,p}^p - \tilde{c}_2|\psi^k|_{1,p}\,.$$

Further from $p \geq 2$ it results

$$|\sigma^k|_{1,p} = |\psi^k|_{1,p} \leq c_3\,. \tag{3.4}$$

Using again Proposition 2.19 (iii), (3.4) entails $\|\sigma^k\|_p \leq c_4$, and hence by (3.3),

$$\|\psi^k\|_p \leq \|\rho^k\|_p + \|\sigma^k\|_p \leq c_5\,. \tag{3.5}$$

Finally (3.4) and (3.5) imply that $\|\psi^k\|_{1,p}$ is bounded in the reflexive space E. Consequently obtain a weak limit point $\hat{\psi} \in E$.
Step 3: *Any weak limit point* $\hat{\psi}$ of (ψ^k) is feasible.
By continuity of the trace map $E \to L^1(\Gamma_c)$, the associated normal component $\hat{u}_n$ satisfies $\hat{u}_n \leq 0$ a.e. on Γ_c.
The presupposed limit behaviour (H_1) of $\tilde{W}$ implies that $\det \nabla\hat{\psi} > 0$ holds a.e. in Ω.
Hence $\hat{\psi} \in V_{\text{ad}}$.
Step 4: *Any weak limit point* $\hat{\psi}$ *of* (ψ^k) *for* $k \to \infty$ *solves* $\mathbb{P}$.

$$X = \{(\psi, A, \delta) | A = \text{adj } \nabla\psi, \delta = \det \nabla\psi\}$$
$$\subset \mathbb{X} = E \times L^q(\Omega, \mathbb{R}^{d\times d}) \times L^r(\Omega)$$

is *not convex*, however it is a sequentially *weakly closed* subset of $\mathbb{X}$, see Proposition 2.17. Hence $(\hat{\psi}, \text{adj } \nabla\hat{\psi}, \det \nabla\hat{\psi}) \in X$.

By polyconvexity, $\tilde{W}$ is a convex function on the open domain $M^d \times M^d \times]0, +\infty[$, hence continuous. By weak lower semicontinuity of convex continuous functionals, conclude

$$I(\hat{\psi}) \leq \liminf_{\nu \to \infty} I(\psi^{(k_\nu)}) \leq \inf_{\psi \in V_{ad}} I(\psi) .$$

Hence, $\hat{\psi}$ is a solution of $\mathbb{P}$.

□

References

1. R.A. Adams, J.J.F. Fournier, *Sobolev Spaces*, 2nd edn. (Elsevier/Academic Press, Amsterdam, 2003)
2. J. Ball, Convexity conditions and existence theorems in nonlinear elasticity. Arch. Rational Mech. Anal. **63**, 337–403 (1977)
3. P.G. Ciarlet, *Mathematical Elasticity. Vol. I. Three-Dimensional Elasticity*. Studies in Mathematics and its Applications, vol. 20 (1988)
4. P.G. Ciarlet, J. Nečas, Unilateral problems in nonlinear, three–dimensional elasticity. Arch. Rational Mech. Anal. **87**, 319–338 (1985)
5. G. Duvaut, J.L. Lions, *Inequalities in Mechanics and Physics* (Springer, Berlin, 1976)
6. Ch. Eck, J. Jarušek, M. Krbec, *Unilateral Contact Problems. Variational Methods and Existence Theorems* (Chapman & Hall/CRC, Boca Raton, 2005)
7. G. Fichera, Boundary value problems of elasticity with unilateral constraints, in *Handbuch der Physik – Encyclopedia of Physics*, ed. by S. Flügge, vol. Band VI a/2 Festkörpermechanik II (Springer, Berlin, 1972), pp. 391–424
8. J. Gwinner, On a frictional unilateral contact problem in nonlinear elasticity - existence and smoothing. Philos. Trans. R. Soc. A **380**(2236), Paper No. 20210356 (2022)
9. N. Kikuchi, J.T. Oden, *Contact Problems in Elasticity. A Study of Variational Inequalities and Finite Element Methods* (SIAM, Philadelphia, 1988)
10. R.W. Ogden, Large deformation isotropic elasticity: on the correlation of theory and experiment for compressible rubberlike solids. Proc. A. **328**, 567–583 (1972)
11. R.T. Rockafellar, *Convex Analysis*, Princeton Landmarks in Mathematics (Princeton University Press, Princeton, 1997), Reprint of the 1970 original

Maximal Noncompactness of Singular Integral Operators on L^p Spaces with Power Weights

Oleksiy Karlovych and Alina Shalukhina

Abstract We consider the singular integral operator $A = aI + bS_\Gamma$ with constant coefficients $a, b \in \mathbb{C}$, where S_Γ is the Cauchy singular integral operator over a curve Γ in the complex plane. We discuss two situations when the operator A is maximally noncompact on the space $L^p(\Gamma, \varrho)$ with $1 < p < \infty$ and a power weight ϱ, that is, when its norm is equal to its essential norm and to its Hausdorff measure of noncompactness. Our results extend two results by Naum Krupnik obtained in earlier 1980s and in 2010.

Keywords Norm · Essential norm · Hausdorff measure of noncompactness · Maximal noncompactness · Cauchy singular integral operator · Power weight

2000 Mathematics Subject Classification 45E05

1 Introduction and the Main Result

For a Banach space X, let $\mathcal{B}(X)$ and $\mathcal{K}(X)$ denote the sets of bounded linear and compact linear operators on X, respectively. The norm of an operator $T \in \mathcal{B}(X)$ is denoted by $\|T\|_{\mathcal{B}(X)}$. The essential norm of $T \in \mathcal{B}(X)$ is defined by

$$\|T\|_{\mathcal{B}(X),\mathrm{e}} := \inf\{\|T - K\|_{\mathcal{B}(X)} : K \in \mathcal{K}(X)\}.$$

O. Karlovych (✉) · A. Shalukhina
Centro de Matemática e Aplicações, Departamento de Matemática, Faculdade de Ciências e Tecnologia, Universidade Nova de Lisboa, Caparica, Portugal
e-mail: oyk@fct.unl.pt; a.shaukhina@campus.fct.unl.pt

R. Duduchava et al. (eds.), *Tbilisi Analysis and PDE Seminar*, Research Perspectives Ghent Analysis and PDE Center 7,
https://doi.org/10.1007/978-3-031-62894-8_9

For a bounded subset Ω of the space X, we denote by $\chi(\Omega)$ the greatest lower bound of the set of numbers r such that Ω can be covered by a finite family of open balls of radius r. For $T \in \mathcal{B}(X)$, set

$$\|T\|_{\mathcal{B}(X),\chi} := \chi(T(B_X)),$$

where B_X denotes the closed unit ball in X. The quantity $\|T\|_{\mathcal{B}(X),\chi}$ is called the Hausdorff measure of noncompactness of the operator T. Consider also the following measure of noncompactness:

$$\|A\|_{\mathcal{B}(X),m} := \inf_{\substack{L \subseteq X \text{ closed linear subspace} \\ \dim(X/L) < \infty}} \left\|A|_L\right\|_{\mathcal{B}(X)},$$

where $A|_L$ denotes the restriction of A to L.

It follows from the definition of the essential norm and [17, inequalities (3.7) and (3.29)] that for every $T \in \mathcal{B}(X)$ one has

$$\frac{1}{2}\|T\|_{\mathcal{B}(X),\chi} \le \|T\|_{\mathcal{B}(X),m} \le 2\|T\|_{\mathcal{B}(X),\chi} \tag{1.1}$$

and

$$\max\{\|T\|_{\mathcal{B}(X),\chi}, \|T\|_{\mathcal{B}(X),m}\} \le \|T\|_{\mathcal{B}(X),\mathrm{e}} \le \|T\|_{\mathcal{B}(X)}. \tag{1.2}$$

Note that there is a typo in [17, inequality (3.7)], where the factor 2 is missing in the right-hand side. The constants 1/2 and 2 in (1.1) are optimal (see [1, Sections 2.5.2 and 2.5.6]). In general, $\|T\|_{\mathcal{B}(X),\mathrm{e}}$ cannot be estimated above by $\|T\|_{\mathcal{B}(X),\chi}$ or $\|T\|_{\mathcal{B}(X),m}$ (see [2]). However, if H is a complex Hilbert space and $T \in \mathcal{B}(H)$, then

$$\|T\|_{\mathcal{B}(H),m} = \|T\|_{\mathcal{B}(H),\chi} = \|T\|_{\mathcal{B}(H),\mathrm{e}} \tag{1.3}$$

(see [23, Theorem 1.1] and also [22, Theorem 4]). We refer to the monographs [1, 3, 4] for the general theory of measures of noncompactness. Following [16, p. 9408] (see also [11]), a bounded operator T on a Banach space X is said to be maximally noncompact if

$$\|T\|_{\mathcal{B}(X),\chi} = \|T\|_{\mathcal{B}(X),\mathrm{e}} = \|T\|_{\mathcal{B}(X)}.$$

Let Γ be a contour in the complex plane. The Cauchy singular integral operator S_Γ of a function $f : \Gamma \to \mathbb{C}$ is defined by

$$(S_\Gamma f)(t) := \frac{1}{\pi i}\int_\Gamma \frac{f(\tau)}{\tau - t}\,d\tau, \quad t \in \Gamma,$$

where the integral is understood in the principal value sense, that is,

$$\int_\Gamma \frac{f(\tau)}{\tau - t}\, d\tau := \lim_{\varepsilon\to 0^+} \int_{\Gamma\setminus\Gamma(t,\varepsilon)} \frac{f(\tau)}{\tau - t}\, d\tau,$$

where $\Gamma(t,\varepsilon) := \{\tau \in \Gamma : |\tau - t| < \varepsilon\}$. A nonnegative locally integrable function w on Γ is said to be a weight. For $1 < p < \infty$, the weighted space $L^p(\Gamma, w)$ consists of all measurable functions $f : \Gamma \to \mathbb{C}$ such that

$$\|f\|_{L^p(\Gamma,w)} := \left(\int_\Gamma |f(t)|^p w(t)\, |dt|\right)^{1/p} < \infty.$$

Equalities of the norm and the essential norm for singular integral operators of the form $A := aI + bS_\Gamma$ with $a, b \in \mathbb{C}$ on weighted Lebesgue spaces were studied in a series of papers by Krupnik and his coauthors. We refer to [6, Section 13.5], [14, Section 6], [15] and to the references therein. Recently, the first author and Shargorodsky obtained the equalities

$$\|A\|_{\mathcal{B}(X(\Gamma)),\chi} = \|A\|_{\mathcal{B}(X(\Gamma)),e} = \|A\|_{\mathcal{B}(X(\Gamma))}$$

in the case when $X(\Gamma)$ is a rearrangement-invariant Banach function space with nontrivial Boyd indices and $\Gamma = \mathbb{T}$ is the unit circle (see [10, Theorem 1.1]) or $\Gamma = \mathbb{R}$ is the real line (see [11, Corollary 5.4 and formula (5.5)]). Thus, the operator A is maximally noncompact on $X(\Gamma)$ in the above sense. Note that rearrangement-invariant Banach function spaces are far-reaching generalisations of Lebesgue spaces. However, Lebesgue spaces with weights are not rearrangement-invariant.

The aim of this paper is to extend two Krupnik's results on $\|A\|_{\mathcal{B}(L^p(\Gamma,\varrho))} = \|A\|_{\mathcal{B}(L^p(\Gamma,\varrho)),e}$, where $1 < p < \infty$ and ϱ is a certain power weight (see [15, Theorem 5.9] and [14, Example 4.2]) in order to include the Hausdorff measure noncompactness $\|A\|_{\mathcal{B}(L^p(\Gamma,\varrho)),\chi}$ into consideration, and to show that the operator $A = aI + bS_\Gamma$ with $a, b \in \mathbb{C}$ is maximally noncompact in the above sense.

Let Γ be a contour in the complex plane consisting of a finite number of circular arcs joining the endpoints -1 and 1, possibly including the segment $[-1, 1]$ as in Fig. 1.

Consider the Khvedelidze weight of the form

$$w(t) := |t-1|^{\alpha_1}|t+1|^{\alpha_2}, \quad t \in \Gamma, \quad \alpha_1, \alpha_2 \in \mathbb{R}.$$

It follows from Khvedelidze's theorem (see, e.g., [5, Ch. 1, Theorem 4.1, Corollary 4.1]) that the Cauchy singular integral operator S_Γ is bounded on $L^p(\Gamma, w)$ if and only if $-1 < \alpha_1, \alpha_2 < p - 1$.

The following theorem addresses the situation $p = 2$ and $\alpha_1 = -\alpha_2 = \beta$.

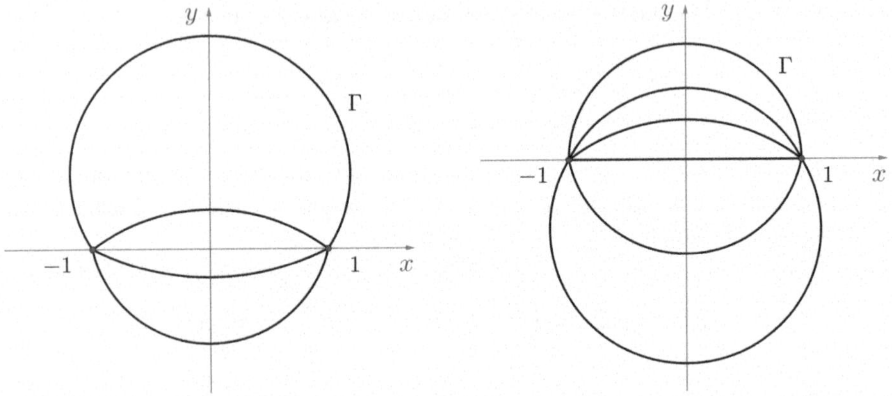

Fig. 1 Examples of a possible contour Γ

Theorem 1.1 *Let* Γ *be a contour in the complex plane consisting of a finite number of circular arcs joining the endpoints* -1 *and* 1*, possibly including the segment* $[-1, 1]$*. Suppose the Khvedelidze weight* ϱ *is given by*

$$\varrho(t) := |t-1|^{\beta}|t+1|^{-\beta}, \quad t \in \Gamma, \quad -1 < \beta < 1.$$

For the singular integral operator $A = aI + bS_{\Gamma}$ *with constant coefficients* $a, b \in \mathbb{C}$*, one has*

$$\|A\|_{\mathcal{B}(L^2(\Gamma,\varrho))} = \|A\|_{\mathcal{B}(L^2(\Gamma,\varrho)),\mathrm{e}} = \|A\|_{\mathcal{B}(L^2(\Gamma,\varrho)),\chi} = \|A\|_{\mathcal{B}(L^2(\Gamma,\varrho)),m}. \tag{1.4}$$

The first equality in (1.4) was announced in [15, Theorem 5.9], its detailed proof is given by the authors in [9, Theorem 1.1] (see also Sect. 2). It seems that the second and the third equalities were not observed earlier. They follow immediately from general equalities (1.3) because $L^2(\Gamma, \varrho)$ is a Hilbert space.

Let us turn our attention to the case of the space $L^p(\mathbb{R}, \varrho)$ with $1 < p < \infty$ and the power weight $\varrho(t) = |t|^{\beta}$ with $-1 < \beta < p-1$. The norm of the operator $S_{\mathbb{R}}$ on this space was computed in 1980 by Krupnik and Verbitsky [21] (see also [14, Theorem 5.2] and [6, Ch. 13, Theorem 3.2]). Further, twenty years later, Hollenbeck and Verbitsky computed the norms of the Riesz projections $(I \pm S_{\mathbb{R}})/2$ on $L^p(\mathbb{R}, \varrho)$ (see [7, Theorem 4.1]).

Theorem 1.2 (Hollenbeck, Krupnik, Verbitsky) *If* $1 < p < \infty$, $1/p + 1/q = 1$, $-1 < \beta < p-1$ *and* $\varrho(t) = |t|^{\beta}$*, then* $S_{\mathbb{R}} \in \mathcal{B}(L^p(\mathbb{R}, \varrho))$ *and*

$$\|S_{\mathbb{R}}\|_{\mathcal{B}(L^p(\mathbb{R},\varrho))} = \nu(p,\beta), \quad \|(I \pm S_{\mathbb{R}})/2\|_{\mathcal{B}(L^p(\mathbb{R},\varrho))} = \mu(p,\beta),$$

where the constant $\nu(p, \beta)$ *is defined for* $2 \le p < \infty$ *by*

$$\nu(p,\beta) := \begin{cases} \cot \dfrac{\pi(p-1-\beta)}{2p} & \text{if } p-2 < \beta < p-1, \\ \cot \dfrac{\pi}{2p} & \text{if } 0 \le \beta \le p-2, \\ \cot \dfrac{\pi(1+\beta)}{2p} & \text{if } -1 < \beta < 0, \end{cases} \tag{1.5}$$

and for $1 < p < 2$ *by*

$$\nu(p,\beta) := \nu(q, (1-q)\beta); \tag{1.6}$$

and the constant $\mu(p, \beta)$ *is defined by*

$$\mu(p,\beta) := \max\left(\frac{1}{\sin\frac{\pi}{p}}, \frac{1}{\sin\frac{\pi(1+\beta)}{p}}\right). \tag{1.7}$$

Our main result is the following.

Theorem 1.3 (Main Result) *Let* $1 < p < \infty$, $-1 < \beta < p-1$, *and* $\varrho(t) = |t|^\beta$. *Then the singular integral operator* $A = aI + bS_{\mathbb{R}}$ *with constant coefficients* $a, b \in \mathbb{C}$ *is maximally noncompact on the space* $L^p(\mathbb{R}, \varrho)$, *that is,*

$$\|A\|_{\mathcal{B}(L^p(\mathbb{R},\varrho))} = \|A\|_{\mathcal{B}(L^p(\mathbb{R},\varrho)),\mathrm{e}} = \|A\|_{\mathcal{B}(L^p(\mathbb{R},\varrho)),\chi}. \tag{1.8}$$

The first equality in (1.8) was observed by Krupnik in [14, Example 4.2]. The second equality in (1.8) is new and its proof is the subject of the paper.

Corollary 1.4 *Let* $1 < p < \infty$, $-1 < \beta < p-1$, *and* $\varrho(t) = |t|^\beta$. *Then*

$$\|S_{\mathbb{R}}\|_{\mathcal{B}(L^p(\mathbb{R},\varrho))} = \|S_{\mathbb{R}}\|_{\mathcal{B}(L^p(\mathbb{R},\varrho)),\mathrm{e}} = \|S_{\mathbb{R}}\|_{\mathcal{B}(L^p(\mathbb{R},\varrho)),\chi} = \nu(p,\beta), \tag{1.9}$$

where $\nu(p, \beta)$ *is defined by* (1.5) *and* (1.6), *and*

$$\begin{aligned} \|(I \pm S_{\mathbb{R}})/2\|_{\mathcal{B}(L^p(\mathbb{R},\varrho))} &= \|(I \pm S_{\mathbb{R}})/2\|_{\mathcal{B}(L^p(\mathbb{R},\varrho)),\mathrm{e}} \\ &= \|(I \pm S_{\mathbb{R}})/2\|_{\mathcal{B}(L^p(\mathbb{R},\varrho)),\chi} = \mu(p,\beta), \end{aligned} \tag{1.10}$$

where $\mu(p, \beta)$ *is defined by* (1.7). *Moreover, if* $p = 2$ *and* $-1 < \beta < 1$, *then*

$$\|S_{\mathbb{R}}\|_{\mathcal{B}(L^2(\mathbb{R},\varrho)),m} = \nu(2,\beta), \quad \|(I \pm S_{\mathbb{R}})/2\|_{\mathcal{B}(L^2(\mathbb{R},\varrho)),m} = \mu(2,\beta). \tag{1.11}$$

Equalities (1.9) and (1.10) follow from Theorems 1.2 and 1.3. Further, equalities (1.11) follow from (1.3) and (1.9) and (1.10) because $L^2(\mathbb{R}, \varrho)$ is a Hilbert space.

The rest of the paper is organised as follows. In Sect. 2, we sketch the proof of Theorem 1.1. Section 3 is devoted to the proof of Theorem 1.3. It is based on Theorem 3.1, whose proof will be published elsewhere.

2 On the Proof of Theorem 1.1

2.1 *Sufficient Conditions for the Equality of the Norm and the Essential Norm on a Banach Space*

Let X be a Banach space and X^* be its dual space. The action of a functional $x^* \in X^*$ on an element $x \in X$ will be denoted by

$$x^*(x) := \langle x, x^* \rangle.$$

The following sufficient condition for the equality of the norm and the essential norm of a bounded operator on a Banach space is due to Krupnik [14, Theorem 4.3].

Theorem 2.1 *Let X be a Banach space and $T \in \mathcal{B}(X)$. Suppose that there exists a sequence $\{L_n\}$ in $\mathcal{B}(X)$ such that*

(a) $\|L_n x\|_X = \|x\|_X$ *for all $x \in X$ and all $n \in \mathbb{N}$;*
(b) $TL_n = L_n T$ *for all $n \in \mathbb{N}$;*
(c) *the sequence $\{L_n\}$ converges weakly to the zero operator on X, that is, $\langle L_n x, x^* \rangle \to 0$ as $n \to \infty$ for all $x \in X$ and $x^* \in X^*$.*

Then $\|T\|_{\mathcal{B}(X)} = \|T\|_{\mathcal{B}(X),e}$.

2.2 *On the Proof of the First Equality in (1.4)*

It follows from [9, Lemma 2.1] that if $\lambda > 1$, then the Möbius transformation

$$t_\lambda(z) = \frac{\lambda z + 1}{z + \lambda}$$

maps Γ onto itself. Further, in view of [9, Lemma 2.2], the operator

$$(L_{\lambda,\mu} f)(z) = \frac{\mu}{z + \lambda} f(t_\lambda(z)), \quad z \in \Gamma,$$

with $\lambda > 1$ and $\mu > 0$, is an isometry on $L^2(\Gamma, \varrho)$ if and only if

$$\mu = (\lambda - 1)^{\frac{1+\beta}{2}} (\lambda + 1)^{\frac{1-\beta}{2}}.$$

Moreover, due to [9, Lemma 2.4], $L_{\lambda,\mu} S_\Gamma = S_\Gamma L_{\lambda,\mu}$. Finally, if $n \ge 2$,

$$\lambda(n) = \frac{n+1}{n-1}, \quad \mu(n) = (\lambda(n)-1)^{\frac{1+\beta}{2}} (\lambda(n)+1)^{\frac{1-\beta}{2}} = \frac{2n^{\frac{1-\beta}{2}}}{n-1},$$

then the sequence $\{L_{\lambda(n),\mu(n)}\}$ converges weakly to the zero operator on $L^2(\Gamma, \varrho)$ as $n \to \infty$ (see [9, Section 2.7]). Thus, all hypotheses in Theorem 2.1 are satisfied. It remains to apply it in order to conclude that $\|A\|_{\mathcal{B}(L^2(\Gamma,\varrho))} = \|A\|_{\mathcal{B}(L^2(\Gamma,\varrho)),\mathrm{e}}$.

3 Proof of the Main Result

3.1 *Sufficient Conditions for the Maximal Noncompactness on Banach Spaces*

The following result is the substitute of Krupnik's Theorem 2.1. Its proof is similar to the proofs of [10, Theorems 1.1–1.2] and [11, Theorem 1.1]. It will be given in [12].

Theorem 3.1 *Let X be a Banach space and $T \in \mathcal{B}(X)$. Suppose that there exist sequences $\{L_n\}$ in $\mathcal{B}(X)$ and $\{R_n^*\}$ in $\mathcal{B}(X^*)$ such that*

(a) *for each $n \in \mathbb{N}$, the operator L_n is a contraction on X;*
(b) *for each $n \in \mathbb{N}$, the operator R_n^* is a contraction on X^*;*
(c) *for each $n \in \mathbb{N}$, one has $TL_n = L_nT$;*
(d) *for all $n \in \mathbb{N}$ and all $x \in X$, $x^* \in X^*$, one has $|\langle L_n x, R_n^* x^*\rangle| \ge |\langle x, x^*\rangle|$;*
(e) *the sequence $\{R_n^*\}$ converges weakly-* to the zero operator on X^*, that is, for all $x \in X$ and all $x^* \in X^*$, one has $\langle x, R_n^* x^*\rangle \to 0$ as $n \to \infty$.*

Then the operator T is maximally noncompact on the space X, that is,

$$\|T\|_{\mathcal{B}(X),\chi} = \|T\|_{\mathcal{B}(X),\mathrm{e}} = \|T\|_{\mathcal{B}(X)}.$$

One can show that hypotheses (a), (b), (d) of Theorem 3.1 yield that the operators L_n and R_n^* are isometries on the spaces X and X^*, respectively, for all $n \in \mathbb{N}$.

The next lemma says that it is enough to verify hypothesis (e) in Theorem 3.1 only on dense subsets of X and X^*.

Lemma 3.2 *Let X be a Banach space and $\{R_n^*\}$ be a uniformly bounded sequence in $\mathcal{B}(X^*)$. Suppose U is a dense set in X and V^* is a dense set in X^*. If $\langle u, R_n^* v^*\rangle \to 0$ as $n \to \infty$ for all $u \in U$ and all $v^* \in V^*$, then $\langle x, R_n^* x^*\rangle \to 0$ as $n \to \infty$ for all $x \in X$ and $x^* \in X^*$.*

The proof of this lemma is standard. It can be developed by analogy with the proof of [20, Lemma 1.4.1(i)].

3.2 Density of Smooth Compactly Supported Functions in $L^p(\mathbb{R}, \varrho)$ and $L^q(\mathbb{R}, \varrho^{1-q})$

Let $C_c^\infty(\mathbb{R})$ denote the set of all infinitely differentiable functions with compact support.

Lemma 3.3 *Let $1 < p < \infty$, $1/p + 1/q = 1$, and $-1 < \beta < p - 1$. If $\varrho(t) = |t|^\beta$, then the set $C_c^\infty(\mathbb{R})$ is dense in $L^p(\mathbb{R}, \varrho)$ and in $L^q(\mathbb{R}, \varrho^{1-q})$.*

Proof It follows from Theorem 1.2 and [8, Theorem 9] that if $-1 < \beta < p - 1$, then weight ϱ belongs to the Muckenhoupt class $A_p(\mathbb{R})$, that is,

$$\sup_{\mathcal{I}\subset\mathbb{R}} \left(\frac{1}{|\mathcal{I}|}\int_{\mathcal{I}} \varrho(t)\,dt\right)\left(\frac{1}{|\mathcal{I}|}\int_{\mathcal{I}} \varrho(t)^{-\frac{1}{p-1}}\,dt\right)^{p-1} < \infty,$$

where the supremum is taken over all interval $\mathcal{I} \subset \mathbb{R}$ of finite length and $|\mathcal{I}|$ denotes the length of $\mathcal{I}$. The inequality $-1 < \beta < p - 1$ implies that

$$-1 = (p-1)(q-1) < (1-q)\beta < q - 1.$$

Hence $-1 < \beta < p - 1$ also implies that $\varrho^{1-q} \in A_q(\mathbb{R})$. Since $\varrho \in A_p(\mathbb{R})$ and $\varrho^{1-q} \in A_q(\mathbb{R})$, the set $C_c^\infty(\mathbb{R})$ is dense in $L^p(\mathbb{R}, \varrho)$ and in $L^q(\mathbb{R}, \varrho^{1-q})$ in view of [18, Lemma 1] (see also [19, Theorem 1.1]). □

The above result can also be extracted from [13, Theorem 4.1].

3.3 Proof of Theorem 1.3

Let $1 < p < \infty$, $1/p + 1/q = 1$ and $-1 < \beta < p - 1$. Then $(L^p(\mathbb{R}, \varrho))^* = L^q(\mathbb{R}, \varrho^{1-q})$ with respect to the pairing

$$(f, g) := \int_{\mathbb{R}} f(t)\overline{g(t)}\,dt.$$

For $f \in L^p(\mathbb{R}, \varrho)$ and $g \in L^q(\mathbb{R}, \varrho^{1-q})$, inspired by [14, Example 4.2], consider the operators

$$(L_n f)(t) := n^{\frac{1+\beta}{p}} f(nt), \quad (R_n^* g)(t) := n^{\frac{1+(1-q)\beta}{q}} g(nt), \quad n \in \mathbb{N}.$$

It is easy to see that

$$\|L_n f\|_{L^p(\mathbb{R},\varrho)} = \left(\int_{\mathbb{R}} n^{1+\beta}|f(nt)|^p|t|^\beta dt\right)^{1/p}$$

$$= n^{\frac{1+\beta}{p}}\left(\int_{\mathbb{R}} |f(x)|^p \left|\frac{x}{n}\right|^\beta \frac{dx}{n}\right)^{1/p} = \|f\|_{L^p(\mathbb{R},\varrho)},$$

$$\|R_n^* g\|_{L^q(\mathbb{R},\varrho^{1-q})} = \left(\int_{\mathbb{R}} n^{1+(1-q)\beta}|g(nt)|^q|t|^{(1-q)\beta} dt\right)^{1/q}$$

$$= n^{\frac{1+(1-q)\beta}{q}}\left(\int_{\mathbb{R}} |g(x)|^q \left|\frac{x}{n}\right|^{(1-q)\beta} \frac{dx}{n}\right)^{1/q} = \|g\|_{L^q(\mathbb{R},\varrho^{1-q})},$$

$$(L_n f, R_n^* g) = n^{\frac{1+\beta}{p}} \cdot n^{\frac{1+(1-q)\beta}{q}} \int_{\mathbb{R}} f(nt)\overline{g(nt)}\, dt$$

$$= n\int_{\mathbb{R}} f(nt)\overline{g(nt)}\, dt = n\int_{\mathbb{R}} f(x)\overline{g(x)}\,\frac{dx}{n} = (f, g).$$

Hence, hypotheses (a), (b), and (d) in Theorem 3.1 are satisfied.

A straightforward calculation also shows that for a.e. $t \in \mathbb{R}$,

$$(S_{\mathbb{R}} L_n f)(t) = \frac{1}{\pi i}\int_{\mathbb{R}} \frac{(L_n f)(\tau)}{\tau - t} d\tau = \frac{n^{\frac{1+\beta}{p}}}{\pi i}\int_{\mathbb{R}} \frac{f(n\tau)}{\tau - t} d\tau$$

$$= \frac{n^{\frac{1+\beta}{p}}}{\pi i}\int_{\mathbb{R}} \frac{f(x)}{x/n - t}\frac{dx}{n} = \frac{n^{\frac{1+\beta}{p}}}{\pi i}\int_{\mathbb{R}} \frac{f(x)}{x - nt} dx = (L_n S_{\mathbb{R}} f)(t).$$

This implies that $AL_n f = L_n A f$, that is, hypothesis (c) in Theorem 3.1 is satisfied.

Let us verify hypothesis (e), that is, that for all $f \in L^p(\mathbb{R}, \varrho)$, $g \in L^q(\mathbb{R}, \varrho^{1-q})$, one has

$$(f, R_n^* g) = n^{\frac{1+(1-q)\beta}{q}} \int_{\mathbb{R}} f(t)\overline{g(nt)}\, dt \to 0 \quad \text{as} \quad n \to \infty. \tag{3.1}$$

In view of Lemmas 3.2 and 3.3, it is enough to prove (3.1) for $f, g \in C_c^\infty(\mathbb{R})$.

So, suppose that $f, g \in C_c^\infty(\mathbb{R})$. Then $M := \sup_{t\in\mathbb{R}} |f(t)| < \infty$, $L := \int_{\mathbb{R}} |g(t)|\, dt < \infty$ and

$$|(f, R_n^* g)| \le n^{\frac{1+(1-q)\beta}{q}} \int_{\mathbb{R}} |f(t)|\,|g(nt)|\, dt \le M n^{\frac{1}{q}-\frac{\beta}{p}} \int_{\mathbb{R}} |g(nt)|\, dt$$

$$= M n^{\frac{1}{q}-1-\frac{\beta}{p}} \int_{\mathbb{R}} |g(x)|\, dx = MLn^{-\frac{1+\beta}{p}}.$$

Since $1+\beta > 0$, the last inequality implies that (3.1) holds for $f, g \in C_c^\infty(\mathbb{R})$. This completes the verification of hypothesis (e).

Thus, all hypotheses (a)–(e) in Theorem 3.1 are satisfied. It remains to apply Theorem 3.1. □

Acknowledgments We would like to thank Eugene Shargorodsky for useful discussions.

This work is funded by national funds through the FCT—Fundação para a Ciência e a Tecnologia, I.P., under the scope of the projects UIDB/00297/2020 and UIDP/00297/2020 (Center for Mathematics and Applications). The second author is funded by national funds through the FCT—Fundação para a Ciência e a Tecnologia, I.P., under the scope of the PhD scholarship UI/BD/154284/2022.

References

1. R.R. Akhmerov, M.I. Kamenskiĭ, A.S. Potapov, A.E. Rodkina, B.N. Sadovskiĭ, *Measures of Noncompactness and Condensing Operators*, volume 55 of Operator Theory: Advances and Applications (Birkhäuser Verlag, Basel, 1992)
2. K. Astala, H.-O. Tylli, On the bounded compact approximation property and measures of noncompactness. J. Funct. Anal. **70**(2), 388–401 (1987)
3. J.M. Ayerbe Toledano, T. Domínguez Benavides, G. López Acedo, *Measures of Noncompactness in Metric Fixed Point Theory*, volume 99 of Operator Theory: Advances and Applications (Birkhäuser Verlag, Basel, 1997)
4. J. Banaś, M. Mursaleen. *Sequence Spaces and Measures of Noncompactness with Applications to Differential and Integral Equations* (Springer, New Delhi, 2014)
5. I. Gohberg, N. Krupnik, *One-dimensional Linear Singular Integral Equations. I. Introduction*, volume 53 of Operator Theory: Advances and Applications (Birkhäuser Verlag, Basel, 1992)
6. I. Gohberg, N. Krupnik, *One-dimensional Linear Singular Integral Equations. II. General Theory and Applications*, volume 54 of Operator Theory: Advances and Applications (Birkhäuser Verlag, Basel, 1992)
7. B. Hollenbeck, I.E. Verbitsky, Best constants for the Riesz projection. J. Funct. Anal. **175**(2), 370–392 (2000)
8. R. Hunt, B. Muckenhoupt, R. Wheeden, Weighted norm inequalities for the conjugate function and Hilbert transform. Trans. Amer. Math. Soc. **176**, 227–251 (1973)
9. O. Karlovych, A. Shalukhina, *Maximal Noncompactness of Singular Integral Operators on L^2 Spaces with some Khvedelidze Weights*, volume 54 of Operator Theory: Advances and Applications (Birkhäuser/Springer, Cham, 2024), pp. 279–295. https://doi.org/10.1007/978-3-031-50613-0_12
10. O. Karlovych, E. Shargorodsky, On the essential norms of singular integral operators with constant coefficients and of the backward shift. Proc. Amer. Math. Soc. Ser. B **9**, 60–70 (2022)
11. O. Karlovych, E. Shargorodsky, Discrete Riesz transforms on rearrangement-invariant Banach sequence spaces and maximally noncompact operators. *Pure and Applied Functional Analysis* **9**(1), 195–210, 2024. https://doi.org/10.1016/j.indag.2023.10.004
12. O. Karlovych, E. Shargorodsky, Maximal noncompactness of dilation-invariant operators (in preparation)
13. V. Kokilashvili, S. Samko, Singular integrals in weighted Lebesgue spaces with variable exponent. Georgian Math. J. **10**(1), 145–156 (2003)
14. N. Krupnik, *Banach Algebras with Symbol and Singular Integral Operators*, volume 26 of Operator Theory: Advances and Applications (Birkhäuser Verlag, Basel, 1987)

15. N. Krupnik, Survey on the best constants in the theory of one-dimensional singular integral operators, in *Topics in Operator Theory. Volume 1. Operators, Matrices and Analytic Functions*, volume 202 of Operator Theory: Advances and Applications (Birkhäuser Verlag, Basel, 2010), pp. 365–393
16. J. Lang, V. Musil, M. Olšák, L. Pick, Maximal non-compactness of Sobolev embeddings. J. Geom. Anal. **31**(9), 9406–9431 (2021)
17. A. Lebow, M. Schechter, Semigroups of operators and measures of noncompactness. J. Funct. Anal. **7**, 1–26 (1971)
18. N. Miller, Weighted Sobolev spaces and pseudodifferential operators with smooth symbols. Trans. Amer. Math. Soc. **269**(1), 91–109 (1982)
19. E. Nakai, N. Tomita, K. Yabuta, Density of the set of all infinitely differentiable functions with compact support in weighted Sobolev spaces. Sci. Math. Jpn. **60**(1), 121–127 (2004)
20. S. Roch, P.A. Santos, B. Silbermann, *Non-commutative Gelfand Theories. A tool-kit for Operator Theorists and Numerical Analysts*. Universitext (Springer-Verlag London, London, 2011)
21. I.E. Verbickiĭ, N.J. Krupnik, Exact constants in theorems on the boundedness of singular operators in L_p spaces with a weight and their application. Mat. Issled. (54), 21–35, 165 (1980) (in Russian)
22. J.R.L. Webb, On seminorms of operators. J. London Math. Soc. (2) **7**, 337–342 (1973)
23. K. Ylinen, Measures of noncompactness for elements of C^*-algebras. Ann. Acad. Sci. Fenn. Ser. A I Math. **6**(1), 131–133 (1981)

A Remark on Piecewise Linear Interpolation of Continuous Fourier Multipliers

Oleksiy Karlovych and Eugene Shargorodsky

Abstract Classical results by de Leeuw (Ann Math (2) 81:364–379, 1965) and Jodeit (Studia Math, 34:215–226, 1970) imply that, for every continuous Fourier multiplier a on $L^p(\mathbb{R})$, $1 < p < \infty$, the piecewise linear function b, satisfying $b(n) = a(n)$ for all $n \in \mathbb{Z}$, is again a Fourier multiplier on $L^p(\mathbb{R})$. We observe that for every $p \in (1, 2) \cup (2, \infty)$, there exists a Lipschitz continuous periodic Fourier multiplier a on $L^p(\mathbb{R})$ and a set $E \subset \mathbb{Z}$ such that the piecewise linear function b with the nodes at the points of E, satisfying $b(n) = a(n)$ for all $n \in E$, fails to be a Fourier multiplier on $L^p(\mathbb{R})$.

Keywords Fourier multiplier · Piecewise linear function · Function of bounded variation

2000 Mathematics Subject Classification 42A45

1 Introduction and the Main Result

Let $S(\mathbb{R})$ denote the class of Schwartz functions (see, e.g., [7, Section 2.2.1]). Given $f \in S(\mathbb{R})$, the Fourier transform of f is defined by

$$\widehat{f}(\xi) = \int_{\mathbb{R}} f(x)e^{-2\pi i x\xi}\,dx.$$

O. Karlovych
Centro de Matemática e Aplicações, Departamento de Matemática, Faculdade de Ciências e Tecnologia, Universidade Nova de Lisboa, Caparica, Portugal
e-mail: oyk@fct.unl.pt

E. Shargorodsky (✉)
Department of Mathematics, King's College London, London, United Kingdom
e-mail: eugene.shargorodsky@kcl.ac.uk

R. Duduchava et al. (eds.), *Tbilisi Analysis and PDE Seminar*, Research Perspectives Ghent Analysis and PDE Center 7,
https://doi.org/10.1007/978-3-031-62894-8_10

The inverse Fourier transform is defined by $\check{f}(x) = \hat{f}(-x)$. By the Plancherel theorem, the operators $f \mapsto \hat{f}$ and $f \mapsto \check{f}$, acting on $S(\mathbb{R})$, extend uniquely to the operators F and F^{-1}, respectively, acting on $L^2(\mathbb{R})$ isometrically, and $FF^{-1} = F^{-1}F = I$.

Given $1 \le p < \infty$, we denote by $M_p(\mathbb{R})$ the set of all functions $a \in L^\infty(\mathbb{R})$ such that

$$\|a\|_{M_p} = \sup\left\{ \frac{\|F^{-1}aFf\|_{L^p(\mathbb{R})}}{\|f\|_{L^p(\mathbb{R})}} : f \in S(\mathbb{R}) \setminus \{0\} \right\} < \infty.$$

The elements of $M_p(\mathbb{R})$ are called Fourier multipliers on $L^p(\mathbb{R})$. It is well known that $M_p(\mathbb{R})$ is a Banach algebra with respect to pointwise operations and the norm $\|\cdot\|_{M_p}$, and that

$$M_1(\mathbb{R}) = \bigcap_{1 \le p < \infty} M_p(\mathbb{R})$$

(see, e.g., [7, Section 2.5.5]).

Let $a \in C(\mathbb{R})$. Take a subset $E \subseteq \mathbb{Z}$ and consider the piecewise linear function b_E with the nodes at E such that $b_E(n) = a(n)$ for all $n \in E$. If E is bounded below (above), b_E is defined to be constant on $(-\infty, \min E]$ (on $[\max E, \infty)$). The function b_E is called the piecewise linear interpolant of a with the nodes at E.

Using the Stechkin theorem (see, e.g., [4, Theorem 2.11] or [5, Theorem 6.2.5]), one can easily show that if $E \subseteq \mathbb{Z}$ and $a \in C(\mathbb{R})$ has bounded variation over $\mathbb{R}$, then

$$b_E \in \bigcap_{1 < p < \infty} M_p(\mathbb{R}). \tag{1.1}$$

(see Lemma 2.1 below).

We are interested in the following question: given an exponent $p \in (1, \infty)$, a set $E \subseteq \mathbb{Z}$, and a function $a \in C(\mathbb{R}) \cap M_p(\mathbb{R})$, can one say that $b_E \in M_p(\mathbb{R})$?

The answer is obviously positive if $p = 2$ and $E \subseteq \mathbb{Z}$, $a \in C(\mathbb{R}) \cap M_p(\mathbb{R})$ are arbitrary, because, as it is well known, $M_2(\mathbb{R}) = L^\infty(\mathbb{R})$ (see, e.g., [7, Theorem 2.5.10]).

The following theorem gives the positive answer for $E = \mathbb{Z}$, an arbitrary exponent $p \in (1, \infty)$, and an arbitrary function $a \in C(\mathbb{R}) \cap M_p(\mathbb{R})$.

Theorem 1.1 *Let $1 < p < \infty$. If $a \in C(\mathbb{R}) \cap M_p(\mathbb{R})$ and $b_{\mathbb{Z}}$ is the piecewise linear interpolant of a with the nodes at $\mathbb{Z}$, then $b_{\mathbb{Z}} \in M_p(\mathbb{R})$ and*

$$\|b_{\mathbb{Z}}\|_{M_p} \le c_p \|a\|_{M_p} \tag{1.2}$$

with a constant c_p that depends only on p.

This theorem can be obtained as a consequence of results by de Leeuw [3] and Jodeit [8] (see Sect. 3 below).

Now let $E \neq \mathbb{Z}$. Since b_E "varies slower" than $b = b_{\mathbb{Z}}$ from Theorem 1.1, one might expect b_E to also belong to $M_p(\mathbb{R})$. However, our main result shows that, in general, b_E may fail to belong to $M_p(\mathbb{R})$.

Theorem 1.2 (Main Result) *Let $p_0 \in (1, 2) \cup (2, \infty)$ and*

$$\mathcal{R}_{p_0} := \{r \in (1, \infty) : |1/r - 1/2| > |1/p_0 - 1/2|\}. \tag{1.3}$$

There exist a Lipschitz continuous periodic function $a \in M_1(\mathbb{R})$ and a set $E \subset \mathbb{Z}$ such that b_E, the piecewise linear interpolant of a with the nodes at E, belongs to $M_{p_0}(\mathbb{R})$, but does not belong to $M_p(\mathbb{R})$ for any $p \in \mathcal{R}_{p_0}$.

The key ingredient in the proof of Theorem 1.2 is a result by Mockenhaupt and Ricker (see [9, Theorem 2] and Theorem 4.2 below).

The paper is organised as follows. In Sect. 2, we prove (1.1). Section 3 is devoted to the proof of Theorem 1.1. Finally, in Sect. 4, we prove our main result, Theorem 1.2.

2 The Case of Continuous Functions of Bounded Variation

The set of all functions $a : \mathbb{R} \to \mathbb{C}$ satisfying

$$V(a) := \sup \sum_{k=1}^{n} |a(x_k) - a(x_{k-1})| < \infty,$$

where the supremum is taken over all partitions of $\mathbb{R}$ of the form

$$-\infty < x_0 < x_1 < \cdots < x_n < +\infty$$

with $n \in \mathbb{N}$, is denoted by $BV(\mathbb{R})$. Then $BV(\mathbb{R})$ is a unital non-separable Banach algebra with respect to pointwise operations and the norm

$$\|a\|_{BV} := \|a\|_{L^\infty(\mathbb{R})} + V(a).$$

Lemma 2.1 *If $a \in C(\mathbb{R}) \cap BV(\mathbb{R})$ and $E \subseteq \mathbb{Z}$, then the piecewise linear function b_E with the nodes at E, satisfying $b_E(n) = a(n)$ for all $n \in E$, belongs to $M_p(\mathbb{R})$ for every $p \in (1, \infty)$ and*

$$\|b_E\|_{M_p} \le C_p \|b_E\|_{BV} \le C_p \|a\|_{BV},$$

where

$$C_p := \begin{cases} \tan(\pi/(2p)) \text{ if } 1 < p \le 2, \\ \cot(\pi/(2p)) \text{ if } 2 < p < \infty. \end{cases}$$

Proof The Stechkin theorem (see, e.g, [4, Theorem 2.11] or [5, Theorem 6.2.5]) says that if $f \in BV(\mathbb{R})$, then

$$\|f\|_{M_p} \le \|S\|_{L^p \to L^p} \|f\|_{BV}, \tag{2.1}$$

where S is the Cauchy singular integral operator given by

$$(Sf)(x) := \frac{1}{\pi i} \lim_{\varepsilon \to 0} \int_{\mathbb{R} \setminus (x-\varepsilon, x+\varepsilon)} \frac{f(t)}{t-x} \, dt.$$

The norm of S was found by Pichorides ([10, Theorem 4.1], see also [6, Ch. 13, Theorem 1.3]):

$$\|S\|_{L^p \to L^p} = C_p. \tag{2.2}$$

On the other hand, the inequality

$$\|b_E\|_{BV} \le \|a\|_{BV} \tag{2.3}$$

follows immediately from the definition of the norm in the algebra $BV(\mathbb{R})$. The desired inequality is an immediate consequence of (2.1)–(2.3). □

3 The Case $E = \mathbb{Z}$

3.1 Algebra $M_p(\mathbb{Z})$

Let and $\mathbb{T} = [0, 1]$. For a complex-valued function $f \in L^1(\mathbb{T})$, the Fourier coefficients are defined by

$$\widehat{f}(n) = \int_{\mathbb{T}} f(x) e^{-2\pi i n x} dx, \quad n \in \mathbb{Z}$$

(cf. [7, Definition 3.1.1]). A trigonometric polynomial on $\mathbb{T}$ is a function of the form

$$p(x) = \sum_{n \in \mathbb{Z}} a(n) e^{2\pi i n x},$$

where $\{a(n)\}_{n\in\mathbb{Z}}$ is finitely supported. The set of all trigonometric polynomials will be denoted by $\mathcal{P}(\mathbb{T})$.

Let $1 \le p < \infty$. Following [7, Definition 4.3.2], a sequence $\{a(n)\}_{n\in\mathbb{Z}} \in \ell^\infty(\mathbb{Z})$ is said to be an $L^p(\mathbb{T})$-multiplier if the operator

$$(L(a)f)(x) = \sum_{n\in\mathbb{Z}} a(n)\widehat{f}(n)e^{2\pi inx}, \quad f \in \mathcal{P}(\mathbb{T}),$$

maps $\mathcal{P}(\mathbb{T})$ into $L^p(\mathbb{T})$ and

$$\|\{a(n)\}_{n\in\mathbb{Z}}\|_{M_p(\mathbb{Z})} := \sup\left\{\frac{\|L(a)f\|_{L^p(\mathbb{T})}}{\|f\|_{L^p(\mathbb{T})}} : f \in \mathcal{P}(\mathbb{T}) \setminus \{0\}\right\} < \infty.$$

The set of all $L^p(\mathbb{T})$-multipliers is denoted by $M_p(\mathbb{Z})$. It is well known that $M_p(\mathbb{Z})$ is a Banach algebra with respect to the pointwise operations and the norm $\|\cdot\|_{M_p(\mathbb{Z})}$ (see, e.g., [7, Remark 4.3.5]).

3.2 The de Leeuw Multiplier Restriction Theorem

The following theorem goes back to de Leeuw [3, Proposition 3.3] (see also [8, Proposition (4.1)] and [7, Theorem 4.3.7]).

Theorem 3.1 *Let $1 < p < \infty$. If $a \in C(\mathbb{R}) \cap M_p(\mathbb{R})$, then $\{a(n)\}_{n\in\mathbb{Z}} \in M_p(\mathbb{Z})$ and*

$$\|\{a(n)\}_{n\in\mathbb{Z}}\|_{M_p(\mathbb{Z})} \le \|a\|_{M_p}. \tag{3.1}$$

3.3 A Generalisation of the Jodeit Multiplier Extension Theorem

The following result by Berkson et al. (see [1, Theorem 3.1]) is a generalisation of Jodeit's theorem [8, Theorem (3.5)].

Theorem 3.2 *Let $1 \le p < \infty$ and $\beta > 0$. Suppose that $\Lambda \in L^\infty(\mathbb{R})$ and* supp $\Lambda \in$ *$[-\beta, \beta]$. For $\{a(n)\}_{n\in\mathbb{Z}} \in \ell^\infty(\mathbb{Z})$, let*

$$(W_\Lambda a)(\xi) := \sum_{n\in\mathbb{Z}} a(n)\Lambda(\xi - n).$$

Then $W_\Lambda a \in M_p(\mathbb{R})$ *whenever* $\{a(n)\}_{n\in\mathbb{Z}} \in M_p(\mathbb{Z})$ *if and only if* $\Lambda \in M_p(\mathbb{R})$. *Moreover, when this is the case, there exists a constant* k_p *independent of* a, Λ *and* β *such that*

$$\|W_\Lambda a\|_{M_p} \le \beta k_p \|\Lambda\|_{M_p} \|\{a(n)\}_{n\in\mathbb{Z}}\|_{M_p(\mathbb{Z})} .$$

3.4 Proof of Theorem 1.1

It follows from Theorem 3.1 that $\{a(n)\}_{n\in\mathbb{Z}} \in M_p(\mathbb{Z})$ and (3.1) holds. Take

$$L(\xi) = \max\{0, 1 - |\xi|\}, \quad \xi \in \mathbb{R}.$$

Then it is easy to see that

$$b_{\mathbb{Z}}(\xi) = \sum_{n\in\mathbb{Z}} a(n)L(\xi - n) = (W_L a)(\xi), \quad \xi \in \mathbb{R}. \tag{3.2}$$

By the Stechkin theorem, $L \in M_p(\mathbb{R})$. It follows from Theorem 3.2 and (3.2) that $b_{\mathbb{Z}} \in M_p(\mathbb{R})$ and

$$\|b_{\mathbb{Z}}\|_{M_p} = \|W_L a\|_{M_p} \le k_p \|L\|_{M_p} \|\{a(n)\}_{n\in\mathbb{Z}}\|_{M_p(\mathbb{Z})} .$$

Combining this inequality with (3.1), we arrive at (1.2) with $c_p = k_p\|L\|_{M_p}$. □

4 The Case $E \ne \mathbb{Z}$

4.1 An "Infinite Coarse Saw" Belonging to $M_1(\mathbb{R})$

The aim of this section is to prove Theorem 1.2. We start with the following lemma dealing with a piecewise linear function, whose graph looks like an "infinite coarse saw".

Lemma 4.1 *Let*

$$R(\xi) := \max\{0, 1 - 2|\xi|\}, \quad a_0(\xi) := \sum_{n\in\mathbb{Z}} R(\xi - n), \quad \xi \in \mathbb{R}. \tag{4.1}$$

Then a_0 *is a Lipschitz continuous periodic function and hence belongs to* $M_1(\mathbb{R})$.

Proof Clearly, R and its periodic extension a_0 are Lipschitz continuous. Hence a has an absolutely convergent Fourier series

$$a_0(\xi) = \sum_{m\in\mathbb{Z}} \widehat{a_0}(m)e^{2\pi im\xi}, \quad \sum_{m\in\mathbb{Z}} |\widehat{a_0}(m)| < \infty$$

(see, e.g., [7, Theorem 3.3.16]). An easy calculation provides a direct proof of this. Indeed, the Fourier series of a_0 is given by

$$a_0(\xi) = \frac{1}{2} + \frac{2}{\pi^2} \sum_{m \text{ odd}} \frac{e^{2\pi im\xi}}{m^2},$$

which is obviously absolutely convergent. Since for every $f \in S(\mathbb{R})$,

$$\left|(F^{-1}a_0Ff)(x)\right| = \left|\sum_{m\in\mathbb{Z}} \widehat{a_0}(m)f(x+m)\right| \leq \sum_{m\in\mathbb{Z}} |\widehat{a_0}(m)|\,|f(x+m)|, \quad x \in \mathbb{R},$$

one has

$$\|a_0\|_{M_1} \leq \sum_{m\in\mathbb{Z}} |\widehat{a_0}(m)| < \infty,$$

which completes the proof. □

4.2 A Theorem by Mockenhaupt and Ricker

For $E \subset \mathbb{R}$ (or $E \subset \mathbb{Z}$), let $\chi_E : \mathbb{R} \to \{0, 1\}$ ($\chi_E : \mathbb{Z} \to \{0, 1\}$, respectively) denote the characteristic (indicator) function of E.

The following result by Mockenhaupt and Ricker is the key ingredient in our proof of Theorem 1.2.

Theorem 4.2 *Let* $p_0 \in (1, 2) \cup (2, \infty)$ *and* $\mathcal{R}_{p_0}$ *be defined by* (1.3).

(a) *There exists a set* $E \subset \mathbb{R}$ *such that* $\chi_E \in M_{p_0}(\mathbb{R})$ *but* $\chi_E \notin M_p(\mathbb{R})$ *for any* $p \in \mathcal{R}_{p_0}$.
(b) *There exists a set* $E \subset \mathbb{Z}$ *such that* $\chi_E \in M_{p_0}(\mathbb{Z})$ *but* $\chi_E \notin M_p(\mathbb{Z})$ *for any* $p \in \mathcal{R}_{p_0}$.

In the case $p > 2$, part (a) is [9, Theorem 2]. For $p > 2$, the proof of part (b) is contained in the proof of that theorem (see [9, p. 64]) and is based on results by Bourgain [2]. If $1 < p < 2$, then the results of parts (a) and (b) are obtained by duality from the corresponding results for $p > 2$.

4.3 Proof of Theorem 1.2

If $p_0 \in (1,2) \cup (2,\infty)$, then by Theorem 4.2(b), there exists a set $E_0 \subset \mathbb{Z}$ such that

$$\chi_{E_0} \in M_{p_0}(\mathbb{Z}), \qquad \chi_{E_0} \notin \bigcup_{p \in \mathcal{R}_{p_0}} M_p(\mathbb{Z}).$$

Let

$$b_0(\xi) := \sum_{n \in E_0} R(\xi - n), \quad \xi \in \mathbb{R},$$

where R is defined by (4.1). Then b_0 is (Lipschitz) continuous, and $b_0(n) = \chi_{E_0}(n)$, $n \in \mathbb{Z}$. Hence $\{b_0(n)\}_{n \in \mathbb{Z}} \notin M_p(\mathbb{Z})$ for every $p \in \mathcal{R}_{p_0}$. Then Theorem 3.1 implies that $b_0 \notin M_p(\mathbb{R})$ for any $p \in \mathcal{R}_{p_0}$.

On the other hand,

$$b_0(\xi) = \sum_{n \in \mathbb{Z}} \chi_{E_0}(n) R(\xi - n) = (W_R \chi_{E_0})(\xi), \quad \xi \in \mathbb{R}. \tag{4.2}$$

By the Stechkin theorem, $R \in M_{p_0}(\mathbb{R})$. It follows from Theorem 3.2 and (4.2) that $b_0 \in M_{p_0}(\mathbb{R})$.

Let

$$E_1 := E_0 \cup \left(\frac{1}{2} + E_0\right) \cup \left(-\frac{1}{2} + E_0\right).$$

It is clear that b_0 is the piecewise linear function with the nodes at E_1 such that

$$b_0(x) = a_0(x), \quad x \in E_1,$$

where a_0 is defined by (4.1) and is a Lipscitz continuous periodic function in $M_1(\mathbb{R})$ (see Lemma 4.1). Finally, let

$$a(x) := a_0(x/2), \quad b_E(x) := b_0(x/2), \quad E := 2E_1 \subset \mathbb{Z}.$$

Then a is periodic and Lipschitz continuous and b_E is the piecewise linear interpolant of a with the nodes at E. It follows from [7, Proposition 2.5.14] that

$$a \in M_1(\mathbb{R}), \quad b_E \in M_{p_0}(\mathbb{R}), \quad b_E \notin \bigcup_{p \in \mathcal{R}_{p_0}} M_p(\mathbb{R}),$$

which completes the proof. □

Acknowledgments This work is funded by national funds through the FCT—Fundação para a Ciência e a Tecnologia, I.P., under the scope of the projects UIDB/00297/2020 and UIDP/00297/2020 (Center for Mathematics and Applications).

References

1. E. Berkson, M. Paluszyński, G. Weiss, Transference couples and their applications to convolution operators and maximal operators, in *Interaction Between Functional Analysis, Harmonic Analysis, and Probability (Columbia, MO, 1994)*, volume 175 of Lecture Notes in Pure and Appl. Math. (Dekker, New York, 1996), pp. 69–84
2. J. Bourgain, Bounded orthogonal systems and the $\Lambda(p)$-set problem. Acta Math. **162**(3–4), 227–245 (1989)
3. K. de Leeuw, On L_p multipliers. Ann. Math. (2) **81**, 364–379 (1965)
4. R. Duduchava, *Integral Equations with Fixed Singularities* (BSB B. G. Teubner Verlagsgesellschaft, Leipzig, 1979)
5. R.E. Edwards, G.I. Gaudry, *Littlewood-Paley and Multiplier Theory*. Ergebnisse der Mathematik und ihrer Grenzgebiete [Results in Mathematics and Related Areas], Band 90 (Springer, Berlin, 1977)
6. I. Gohberg, N. Krupnik, *One-dimensional Linear Singular Integral Equations. Vol. II. General Theory and Applications*, volume 54 of Operator Theory: Advances and Applications (Birkhäuser Verlag, Basel, 1992)
7. L. Grafakos, *Classical Fourier Analysis*, volume 249 of Graduate Texts in Mathematics, 3rd edn. (Springer, New York, 2014)
8. M. Jodeit Jr., Restrictions and extensions of Fourier multipliers. Studia Math. **34**, 215–226 (1970)
9. G. Mockenhaupt, W.J. Ricker, Idempotent multipliers for $L^p(\mathbf{R})$. Arch. Math. (Basel) **74**(1), 61–65 (2000)
10. S.K. Pichorides, On the best values of the constants in the theorems of M. Riesz, Zygmund and Kolmogorov. Studia Math. **44**, 165–179 (errata insert) (1972)

Banach Algebras of Convolution Type Operators with PQC Data

Yuri Karlovich

Abstract Banach algebras $\mathfrak{A}_{\mathbb{R},p,w}$ of convolution type operators with piecewise quasicontinuous data that have piecewise slowly oscillating asymptotics at infinity are studied on weighted Lebesgue spaces $L^p(\mathbb{R}, w)$ with $p \in (1, \infty)$ and a subclass of Muckenhoupt weights w. A Fredholm symbol calculus for the algebra $\mathfrak{A}_{\mathbb{R},p,w}$ is constructed and a Fredholm criterion for the operators $A \in \mathfrak{A}_{\mathbb{R},p,w}$ is established.

Keywords Convolution type operators · Piecewise slowly oscillating function · Piecewise quasicontinuous function · Banach algebra · Fredholm symbol calculus · Fredholmness

1 Outline

Given $p \in (1, \infty)$ and a Muckenhoupt weight $w \in A_p(\mathbb{R})$, let $\mathcal{B}_{p,w}$ be the Banach algebra of all bounded linear operators acting on the weighted Lebesgue space $L^p(\mathbb{R}, w)$. Let $PQC_{\mathbb{R}}$ be the C^*-algebra of all piecewise quasicontinuous functions on the real line $\mathbb{R}$ that are equivalent to piecewise slowly oscillating functions at infinity and let $PQC_{\mathbb{R},p,w}$ be the corresponding Banach algebra of Fourier multipliers on the space $L^p(\mathbb{R}, w)$. For a subclass of weights w, the Banach algebras $\mathfrak{A}_{\mathbb{R},p,w} \subset \mathcal{B}_{p,w}$ generated by all multiplication operators aI with $a \in PQC_{\mathbb{R}}$ and by all convolution operators $W^0(b)$ with $b \in PQC_{\mathbb{R},p,w}$ are studied. First, a Fredholm symbol calculus is constructed for the Banach algebra $\mathcal{Z}_{\mathbb{R},p,w} \subset \mathfrak{A}_{\mathbb{R},p,w}$ generated by the operators $aW^0(b)$ with quasicontinuous data functions $a \in QC_{\mathbb{R}}$ and $b \in QC_{\mathbb{R},p,w}$ that are equivalent to some slowly oscillating functions at infinity. Then a Fredholm symbol calculus for the Banach algebra $\mathfrak{A}_{\mathbb{R},p,w}$ is constructed and a Fredholm criterion for the operators $A \in \mathfrak{A}_{\mathbb{R},p,w}$ is established by applying the

Y. Karlovich (✉)
Centro de Investigación en Ciencias, Universidad Autónoma del Estado de Morelos, Cuernavaca, Mexico
e-mail: karlovich@uaem.mx

R. Duduchava et al. (eds.), *Tbilisi Analysis and PDE Seminar*,
Research Perspectives Ghent Analysis and PDE Center 7,
https://doi.org/10.1007/978-3-031-62894-8_11

Allan-Douglas local principle, limit operators techniques and the two idempotents theorem. The results are partially based on a joint work with C.A. Fernandes and A.Yu. Karlovich.

2 Data Functions

Muckenhout Weights and Fourier Multipliers A measurable function $w : \mathbb{R} \to [0, \infty]$ is called a weight if the preimage $w^{-1}(\{0, \infty\})$ of the set $\{0, \infty\}$ has measure zero. For $p \in (1, \infty)$, a weight w belongs to the *Muckenhoupt class* $A_p(\mathbb{R})$ (see, e.g., [3]) if

$$c_{p,w} := \sup_I \left(\frac{1}{|I|}\int_I w^p(x)dx\right)^{1/p}\left(\frac{1}{|I|}\int_I w^{-q}(x)dx\right)^{1/q} < \infty,$$

where $1/p + 1/q = 1$, and supremum is taken over all intervals $I \subset \mathbb{R}$ of finite length $|I|$.

Given $p \in (1, \infty)$ and $w \in A_p(\mathbb{R})$, we consider the weighted Lebesgue space $L^p(\mathbb{R}, w)$ equipped with norm

$$\|f\|_{L^p(\mathbb{R},w)} := \left(\int_{\mathbb{R}} |f(x)|^p w^p(x)dx\right)^{1/p}.$$

The Banach Algebra $M_{p,w}$ Let $\mathcal{F} : L^2(\mathbb{R}) \to L^2(\mathbb{R})$ denote the *Fourier transform*,

$$(\mathcal{F}f)(x) := \int_{\mathbb{R}} f(t)e^{itx}dt, \quad x \in \mathbb{R}.$$

A function $a \in L^\infty(\mathbb{R})$ is called a *Fourier multiplier* on $L^p(\mathbb{R}, w)$ if the convolution operator $W^0(a) := \mathcal{F}^{-1}a\mathcal{F}$ maps the dense subset $L^2(\mathbb{R}) \cap L^p(\mathbb{R}, w)$ of $L^p(\mathbb{R}, w)$ into itself and extends to a bounded linear operator on $L^p(\mathbb{R}, w)$. Let $M_{p,w}$ stand for the unital Banach algebra of all Fourier multipliers on $L^p(\mathbb{R}, w)$ equipped with pointwise operations and the norm $\|a\|_{M_{p,w}} := \|W^0(a)\|_{\mathcal{B}(L^p(\mathbb{R},w))}$.

Functions $a, c \in L^\infty(\mathbb{R})$ are called equivalent at ∞ ($a \overset{L^\infty}{\sim} c$) if

$$\lim_{N\to\infty} \|a - c\|_{L^\infty(\mathbb{R}\setminus[-N,N])} = 0. \tag{2.1}$$

Let χ_Y denote the characteristic function of a set $Y \subset \mathbb{R}$. Given $p \in (1, \infty)$ and $w \in A_p(\mathbb{R})$, two functions $b, d \in M_{p,w}$ are called equivalent at ∞ ($b \overset{M_{p,w}}{\sim} d$) if

$$\lim_{N\to\infty} \|\chi_{\mathbb{R}\setminus[-N,N]}(b - d)\|_{M_{p,w}} = 0. \tag{2.2}$$

Piecewise Slowly Oscillating Functions on $\mathbb{R}$ Let $\dot{\mathbb{R}} := \mathbb{R} \cup \{\infty\}$ and $\overline{\mathbb{R}} := [-\infty, +\infty]$. For a bounded measurable function $f : \dot{\mathbb{R}} \to \mathbb{C}$ and a set $I \subset \dot{\mathbb{R}}$, let

$$\operatorname{osc}(f, I) = \operatorname{ess\,sup}\big\{|f(t) - f(s)| : t, s \in I\big\}.$$

We say that a function $f \in L^\infty(\mathbb{R})$ is *slowly oscillating at the point* $\lambda \in \dot{\mathbb{R}}$ if for every (equivalently, some) $r \in (0, 1)$,

$$\lim_{x \to +0} \operatorname{osc}\big(f, \lambda + ([-x, -rx] \cup [rx, x])\big) = 0 \text{ if } \lambda \in \mathbb{R},$$

$$\lim_{x \to +\infty} \operatorname{osc}\big(f, [-x, -rx] \cup [rx, x]\big) = 0 \text{ if } \lambda = \infty.$$

For every $\lambda \in \dot{\mathbb{R}}$, we consider the C^*-algebra

$$SO_\lambda := \big\{f \in L^\infty(\mathbb{R}) \cap C(\dot{\mathbb{R}} \setminus \{\lambda\}) : f \text{ slowly oscillates at } \lambda\big\} \subset L^\infty(\mathbb{R}).$$

Let $SO^\diamond$ be the minimal C^*-subalgebra of $L^\infty(\mathbb{R})$ that contains all the C^*-algebras SO_λ with $\lambda \in \dot{\mathbb{R}}$. In particular, $SO^\diamond \supset C(\dot{\mathbb{R}})$. Let PC stand for the C^*-algebra of piecewise continuous functions $f : \mathbb{R} \to \mathbb{C}$. Let $PSO^\diamond = \operatorname{alg}(PC, SO^\diamond) \subset L^\infty(\mathbb{R})$ be the C^*-algebra generated by the C^*-algebras PC and $SO^\diamond$.

Piecewise Quasicontinuous Functions on $\mathbb{R}$ Let H^∞ be the closed subalgebra of $L^\infty(\mathbb{R})$ that consists of all functions being non-tangential limits on $\mathbb{R}$ of bounded analytic functions on the upper half-plane. The C^*-algebra QC of quasicontinuous functions on $\dot{\mathbb{R}}$ is defined by

$$QC := (H^\infty + C(\dot{\mathbb{R}})) \cap (\overline{H^\infty} + C(\dot{\mathbb{R}}))$$

Let $PQC = \operatorname{alg}(PC, QC)$ denote the C^*-algebra of piecewise quasicontinuous functions generated in $L^\infty(\mathbb{R})$ by the C^*-algebras PC and QC. One can define $M_{p,w}$-analogues $SO^\diamond_{p,w}$ and $QC_{p,w}$ of $SO^\diamond$ and QC, respectively, that are Banach algebras contained in $M_{p,w}$. Let $PC_{p,w}$ denote the closure in $M_{p,w}$ of the set of all functions $a \in PC$ of bounded total variation, let $PSO^\diamond_{p,w}$ be the Banach subalgebra of $M_{p,w}$ generated by the Banach algebras $PC_{p,w}$ and $SO^\diamond_{p,w}$, and let $PQC_{p,w}$ denote the Banach subalgebra of $M_{p,w}$ generated by the Banach algebras $PC_{p,w}$ and $QC_{p,w}$.

Piecewise Quasicontinuous Functions with a Specific Behavior at ∞ Using (2.1), we consider the following C^*-subalgebras of the C^*-algebras QC and PQC, respectively:

$$\begin{aligned} QC_{\mathbb{R}} &:= \big\{a \in QC : a \overset{L^\infty}{\sim} c,\ c \in SO^\diamond\big\}, \\ PQC_{\mathbb{R}} &:= \big\{a \in PQC : a \overset{L^\infty}{\sim} c,\ c \in PSO^\diamond\big\}. \end{aligned} \tag{2.3}$$

Given $p \in (1, \infty)$ and $w \in A_p(\mathbb{R})$, and applying (2.2), we also define the following Banach subalgebras of $QC_{p,w}$ and $PQC_{p,w}$, respectively:

$$\begin{aligned} QC_{\mathbb{R},p,w} &:= \big\{b \in QC_{p,w} : b \overset{M_{p,w}}{\sim} d,\ d \in SO^{\diamond}_{p,w}\big\}, \\ PQC_{\mathbb{R},p,w} &:= \big\{b \in PQC_{p,w} : a \overset{M_{p,w}}{\sim} d,\ d \in PSO^{\diamond}_{p,w}\big\}. \end{aligned} \tag{2.4}$$

3 Banach Algebra $\mathfrak{A}_{\mathbb{R},p,w}$

Let $\mathcal{B}(X)$ be the Banach algebra of all bounded linear operators acting on a Banach space X, and let $\mathcal{K}(X)$ be the ideal of all compact operators in $\mathcal{B}(X)$. A weight $w = e^v \in A_p(\mathbb{R})$ is called *a slowly oscillating weight*, $w \in SO$, if for every $\lambda \in \dot{\mathbb{R}}$ there exists a function $\sigma_\lambda \in SO_\lambda$ such that

$$\lim_{x\to\lambda}[(x-\lambda)v'(x) - \sigma_\lambda(x)] = 0 \ \ (\lambda \in \mathbb{R}), \quad \lim_{x\to\infty}[xv'(x) - \sigma_\infty(x)] = 0.$$

Let $\mathcal{B}_{p,w} := \mathcal{B}(L^p(\mathbb{R}, w))$ for $p \in (1, \infty)$ and $w \in A_p(\mathbb{R})$. Consider the Banach algebra

$$\mathfrak{A}_{\mathbb{R},p,w} := \text{alg}\,\big\{aI,\, W^0(b) : \ a \in PQC_{\mathbb{R}},\ b \in PQC_{\mathbb{R},p,w}\big\} \subset \mathcal{B}_{p,w}$$

generated by all multiplication operators aI with $a \in PQC_{\mathbb{R}}$ given by (2.3) and all convolution operators $W^0(b)$ with $b \in PQC_{\mathbb{R},p,w}$ given by (2.4), where $w \in SO$.

For $p \in (1, \infty)$, the Fredholm criteria and index formulas for the operators in the Banach algebra

$$\mathfrak{A}_p(PC) := \text{alg}\,\big\{aI,\, W^0(b) : \ a \in PC,\ b \in PC_p\big\} \subset \mathcal{B}_p$$

were obtained by R.V. Duduchava (see [7–9]). For weighted Lebesgue spaces $L^p(\mathbb{R}, w)$, the Banach algebra $\mathfrak{A}_{p,w}(PC) \subset \mathcal{B}_{p,w}$ was studied by R. Schneider [16] and by S. Roch and B. Silbermann (see [14]) for power weights w, and by A. Böttcher and I.M. Spitkovsky [5, 6] for general Muckenhoupt weights w.

The Fredholmness of operators in the Banach algebra

$$\mathfrak{A}_{p,w}(PSO^{\diamond}) := \text{alg}\,\big\{aI,\, W^0(b) : \ a \in PSO^{\diamond},\ b \in PSO^{\diamond}_{p,w}\big\} \subset \mathcal{B}_{p,w}$$

was studied in [1, 2] in case $w = 1$, $a \in \text{alg}(SO_\infty, PC)$ and $b \in \text{alg}(SO_{\infty,p}, PC_p)$, and in [11, 12] for $a \in PSO^{\diamond}$, $b \in PSO^{\diamond}_{p,w}$ and $w \in SO$.

For $p \in (1, \infty)$, $w \in SO \subset A_p(\mathbb{R})$, we also consider the Banach algebra

$$\mathcal{Z}_{\mathbb{R},p,w} := \text{alg}\,\big\{aW^0(b) : \ a \in QC_{\mathbb{R}},\ b \in QC_{\mathbb{R},p,w}\big\} \subset \mathfrak{A}_{\mathbb{R},p,w}$$

generated by the operators $aW^0(b)$ with $a \in QC_{\mathbb{R}}$, $b \in QC_{\mathbb{R},p,w}$. If $a \in QC_{\mathbb{R}}$ and $b \in PQC_{\mathbb{R},p,w}$, or $a \in PQC_{\mathbb{R}}$ and $b \in QC_{\mathbb{R},p,w}$, then the commutators $[aI, W^0(b)]$ are compact on $L^p(\mathbb{R}, w)$.

Maximal Ideal Spaces and Fibers For a commutative unital Banach algebra $\mathcal{A}$, let $M(\mathcal{A})$ be the maximal ideal space of $\mathcal{A}$. Identifying $\lambda \in \dot{\mathbb{R}}$ with evaluation functionals δ_λ on $\dot{\mathbb{R}}$ gives $M(SO^\diamond) = \bigcup_{\lambda \in \dot{\mathbb{R}}} M_\lambda(SO^\diamond)$, where $M_\lambda(SO^\diamond) = \{\xi \in M(SO^\diamond) : \xi|_{C(\dot{\mathbb{R}})} = \lambda\}$ are fibers of $M(SO^\diamond)$ over $\lambda \in \dot{\mathbb{R}}$. Similarly, $M(QC) = \bigcup_{\lambda \in \dot{\mathbb{R}}} M_\lambda(QC)$, where $M_\lambda(QC) := \{\xi \in M(QC) : \xi|_{C(\dot{\mathbb{R}})} = \lambda\}$ are fibers of $M(QC)$ over $\lambda \in \dot{\mathbb{R}}$.

By [15], for every $\lambda \in \dot{\mathbb{R}}$, let $M_\lambda^\pm(QC)$ be the set of all $\xi \in M_\lambda(QC)$ such that $\xi(a) = 0$ if $a \in QC$ and $\limsup_{\tau \to \lambda^\pm} |a(\tau)| = 0$. Then $M_\lambda^0(QC) := M_\lambda^+(QC) \cap M_\lambda^-(QC)$, $\widetilde{M}_\lambda^\pm(QC) := M_\lambda^\pm(QC) \setminus M_\lambda^0(QC)$, and $M_\lambda^+(QC) \cup M_\lambda^-(QC) = M_\lambda(QC)$.

Further, let $M(PSO^\diamond) = \bigcup_{\lambda \in \dot{\mathbb{R}}} M_\lambda(PSO^\diamond)$, $M_\lambda(PSO^\diamond) = M_\lambda(SO^\diamond) \times \{0, 1\}$. But $M(PQC) = \bigcup_{\lambda \in \dot{\mathbb{R}}} M_\lambda(PQC) \subset M(QC) \times \{0, 1\}$. By [11], for every $\lambda \in \dot{\mathbb{R}}$, $M_\lambda(SO^\diamond_{p,w}) = M_\lambda(SO^\diamond)$. One can prove that $M_\lambda(QC_{\mathbb{R},p,w}) = M_\lambda(QC_{\mathbb{R}}) = \bigcup_{\lambda \in \mathbb{R}} M_\lambda(QC) \cup M_\infty(SO^\diamond)$.

4 Banach Algebra $\mathcal{Z}^\pi_{\mathbb{R},p,w}$

Defining $\mathcal{K}_{p,w} := \mathcal{K}(L^p(\mathbb{R}, w))$, let us study the invertibility of cosets in the commutative quotient Banach algebra $\mathcal{Z}^\pi_{\mathbb{R},p,w} := \mathcal{Z}_{\mathbb{R},p,w}/\mathcal{K}_{p,w}$, which is a central subalgebra of the quotient Banach algebra $\mathfrak{A}^\pi_{\mathbb{R},p,w} := \mathfrak{A}_{\mathbb{R},p,w}/\mathcal{K}_{p,w}$.

Theorem 4.1 *Let $p \in (1, \infty)$ and $w = e^v \in A_p(\mathbb{R})$ be a slowly oscillating weight. The maximal ideal space $M(\mathcal{Z}^\pi_{\mathbb{R},p,w})$ of the commutative quotient Banach algebra $\mathcal{Z}^\pi_{\mathbb{R},p,w}$ is homeomorphic to the set*

$$\begin{aligned}\Omega :=& \Big(\bigcup_{\lambda \in \mathbb{R}} M_\lambda(QC) \times M_\infty(SO^\diamond)\Big) \cup \Big(M_\infty(SO^\diamond) \times \bigcup_{\lambda \in \mathbb{R}} M_\lambda(QC)\Big) \\ &\cup \big(M_\infty(SO^\diamond) \times M_\infty(SO^\diamond)\big)\end{aligned} \tag{4.1}$$

equipped with topology induced by the product topology of

$$\Big(\bigcup_{\lambda \in \mathbb{R}} M_\lambda(QC) \cup M_\infty(SO^\diamond)\Big) \times \Big(\bigcup_{\lambda \in \mathbb{R}} M_\lambda(QC) \cup M_\infty(SO^\diamond)\Big).$$

The Gelfand transform $\Gamma : \mathcal{Z}^{\pi}_{\mathbb{R},p,w} \to C(\Omega)$ *is defined on the generators* $[aW^0(b)]^{\pi}$ *of* $\mathcal{Z}^{\pi}_{\mathbb{R},p,w}$ *(*$a \in QC_{\mathbb{R}}$ *and* $b \in QC_{\mathbb{R},p,w}$*) by*

$$\big[\Gamma([aW^0(b)]^{\pi})\big](\xi,\eta) = a(\xi)b(\eta) \ \textit{for all} \ (\xi,\eta) \in \Omega.$$

Let the Banach algebra $\Lambda = \Lambda_{p,w} \subset \mathcal{B}_{p,w}$ consist of all operators in $\mathcal{B}_{p,w}$ that commute modulo compact operators with every operator $A \in \mathcal{Z}_{\mathbb{R},p,w}$. The quotient Banach algebra $\Lambda^{\pi} = \Lambda/\mathcal{K}_{p,w}$ consists of the cosets $A^{\pi} := A + \mathcal{K}_{p,w}$ $(A \in \Lambda)$, and $\mathfrak{A}^{\pi}_{\mathbb{R},p,w} \subset \Lambda^{\pi}$. For every pair $(\xi,\eta) \in \Omega$, the maximal ideal $\mathcal{I}^{\pi}_{\xi,\eta}$ of $\mathcal{Z}^{\pi}_{\mathbb{R},p,w}$ is generated by the maximal ideals

$$\begin{aligned} \mathcal{I}^{\pi}_{1,\xi} &:= \{[aI]^{\pi} : a \in QC_{\mathbb{R}},\ a(\xi) = 0\} \ \text{and} \\ \mathcal{I}^{\pi}_{2,\eta} &:= \{[W^0(b)]^{\pi} : b \in QC_{\mathbb{R},p,w},\ b(\eta) = 0\} \end{aligned} \tag{4.2}$$

of the commutative Banach algebras $\mathcal{A}^{\pi}_1 := \big\{[aI]^{\pi} : a \in QC_{\mathbb{R}}\big\} \subset \mathcal{B}^{\pi}_{p,w}$ and $\mathcal{A}^{\pi}_2 := \big\{[W^0(b)]^{\pi} : b \in QC_{\mathbb{R},p,w}\big\} \subset \mathcal{B}^{\pi}_{p,w}$, respectively. For each $(\xi,\eta) \in \Omega$, let $J^{\pi}_{\xi,\eta}$ be the smallest closed two-sided ideal of Λ^{π} that contains the maximal ideal $\mathcal{I}^{\pi}_{\xi,\eta}$ of $\mathcal{Z}^{\pi}_{\mathbb{R},p,w}$. Consider the quotient Banach algebra $\Lambda^{\pi}_{\xi,\eta} := \Lambda^{\pi}/J^{\pi}_{\xi,\eta}$ consisting of the cosets $A^{\pi}_{\xi,\eta} := A^{\pi} + J^{\pi}_{\xi,\eta}$ for all $A^{\pi} \in \Lambda^{\pi}$. Let $\mathcal{A}^{\pi}_{\mathbb{R},\xi,\eta}$ be the smallest closed subalgebra of $\Lambda^{\pi}_{\xi,\eta}$ that contains the set $\{A^{\pi}_{\xi,\eta} : A^{\pi} \in \mathfrak{A}^{\pi}_{\mathbb{R},p,w}\}$.

By the Allan-Douglas local principle (see [4, Theorem 1.35]), we obtain the following.

Theorem 4.2 *An operator* $A \in \mathfrak{A}_{\mathbb{R},p,w}$ *is Fredholm on the space* $L^p(\mathbb{R}, w)$ *if and only if for every pair* $(\xi,\eta) \in \Omega$ *the coset*$A^{\pi}_{\xi,\eta}$ *is invertible in the Banach algebra* $\Lambda^{\pi}_{\xi,\eta}$ *(equivalently, in the Banach algebra* $\mathcal{A}^{\pi}_{\mathbb{R},\xi,\eta}$*).*

The Closed Two-Sided Ideal $\mathcal{I}^{\pi}_{\xi,\eta,\mu}$ Given $(\lambda,\tau) \in \widehat{\Omega}_0 := (\mathbb{R}\times\{\infty\})\cup(\{\infty\}\times\mathbb{R})$, we consider the commutative Banach algebra $\mathcal{D}^{\pi}_{\lambda,\tau} \subset \mathcal{B}^{\pi}_{p,w}$ generated by the cosets I^{π} and

$$[\widehat{X}_{\lambda,\tau}]^{\pi} := \begin{cases} [I - (\chi^+_{\lambda} I - W^0(\chi^-_0))^2]^{\pi} & \text{if } (\lambda,\tau) \in \mathbb{R}\times\{\infty\}, \\ [I - (\chi^-_0 I - W^0(\chi^+_{\tau}))^2]^{\pi} & \text{if } (\lambda,\tau) \in \{\infty\}\times\mathbb{R}. \end{cases}$$

Fix $(\lambda,\tau) \in \widehat{\Omega}_0$ and consider the commutative Banach algebra $\mathfrak{N}^{\pi}_{\lambda,\tau}$ generated in $\mathcal{B}^{\pi}_{p,w}$ by the commutative Banach algebras $\mathcal{A}^{\pi}_1$, $\mathcal{A}^{\pi}_2$ and $\mathcal{D}^{\pi}_{\lambda,\tau}$. Let $\mathcal{N}_{\lambda,\tau}$ be the spectrum of the coset $[\widehat{X}_{\lambda,\tau}]^{\pi}$. For each $(\xi,\eta,\mu) \in M_{\lambda}(QC_{\mathbb{R}})\times M_{\tau}(QC_{\mathbb{R}})\times\mathcal{N}_{\lambda,\tau}$, we consider the closed two-sided ideal $\mathcal{I}^{\pi}_{\xi,\eta,\mu}$ of the commutative Banach algebra $\mathfrak{N}^{\pi}_{\lambda,\tau}$, which is generated by the maximal ideals (4.2) and by the maximal ideal $\mathcal{I}^{\pi}_{3,\mu}$ of $\mathcal{D}^{\pi}_{\lambda,\tau}$ that contains the element $[\widehat{X}_{\lambda,\tau} - \mu I]^{\pi}$.

Let $X^{\pi}_{\xi,\eta} := [\widehat{X}_{\lambda,\tau}]^{\pi} + J^{\pi}_{\xi,\eta}$ for all $(\xi,\eta) \in M_{\lambda}(QC_{\mathbb{R}}) \times M_{\tau}(QC_{\mathbb{R}})$ and all $(\lambda,\tau) \in \widehat{\Omega}_0$.

5 The Study

For $\lambda \in \mathbb{R}$, let χ_λ^- and χ_λ^+ be the characteristic functions of the intervals $(-\infty, \lambda)$ and $(\lambda, +\infty)$, respectively, and let $\chi_\pm := \chi_0^\pm$. For each $g \in PSO_{p,w}^\diamond$ and each $\zeta \in M_\infty(SO^\diamond)$, one can define the values $g(\zeta^-) = g(\zeta, 0)$ and $g(\zeta^+) = g(\zeta, 1)$. To every $a \in PQC_{\mathbb{R}}$, every $b \in PQC_{\mathbb{R},p,w}$ and every $\zeta \in M_\infty(SO^\diamond)$, we assign the functions $\widetilde{a}_\zeta \in C(\overline{\mathbb{R}})$ and $\widetilde{b}_\zeta \in C_{p,w}(\overline{\mathbb{R}})$ by

$$\widetilde{a}_\zeta := a(\zeta^+)u_- + a(\zeta^-)u_+, \quad \widetilde{b}_\zeta := b(\zeta^+)u_- + b(\zeta^-)u_+ \tag{5.1}$$

where the values $a(\zeta^\pm) = c(\zeta^\pm)$ and $b(\zeta^\pm) = d(\zeta^\pm)$ are given for the associated at ∞ functions $c \in PSO^\diamond$ and $d \in PSO_{p,w}^\diamond$, respectively, and the functions $u_\pm \in C_{p,w}(\overline{\mathbb{R}})$ are given by $u_\pm(x) = (1 \pm \tanh x)/2$ for $x \in \overline{\mathbb{R}}$.

In what follows, according to [15], for $a \in PQC_{\mathbb{R}}$ and $b \in PQC_{\mathbb{R},p,w}$, we write

$$\begin{aligned}
&a(\xi^-) = a(\xi, 0) \text{ and } a(\xi^+) = a(\xi, 1) \text{ if } \xi \in M_\lambda^0(QC),\ \lambda \in \mathbb{R},\\
&a(\xi^-) = a(\xi, 0) \text{ if } (\xi, 0) \in M(PQC) \text{ for } \xi \in \widetilde{M}_\lambda^-(QC),\ \lambda \in \mathbb{R},\\
&a(\xi^+) = a(\xi, 1) \text{ if } (\xi, 1) \in M(PQC) \text{ for } \xi \in \widetilde{M}_\lambda^+(QC),\ \lambda \in \mathbb{R},\\
&b(\eta^-) = b(\eta, 0) \text{ and } b(\eta^+) = b(\eta, 1) \text{ if } \eta \in M_\tau^0(QC),\ \tau \in \mathbb{R},\\
&b(\eta^-) = b(\eta, 0) \text{ if } (\eta, 0) \in M(PQC) \text{ for } \eta \in \widetilde{M}_\tau^-(QC),\ \tau \in \mathbb{R},\\
&b(\eta^+) = b(\eta, 1) \text{ if } (\eta, 1) \in M(PQC) \text{ for } \eta \in \widetilde{M}_\tau^+(QC),\ \tau \in \mathbb{R}.
\end{aligned}$$

Theorem 5.1 *Let Ω be given by (4.1). For every $(\xi, \eta) \in \Omega$ and every $\lambda, \tau \in \mathbb{R}$, the mapping $\delta_{\xi,\eta} : A \mapsto A_{\xi,\eta}^\pi$ defined on the generators aI ($a \in PQC_{\mathbb{R}}$) and $W^0(b)$ ($b \in PQC_{\mathbb{R},p,w}$) of the Banach algebra $\mathfrak{A}_{\mathbb{R},p,w}$ by the formulas*

$$\delta_{\xi,\eta}(aI) := \begin{cases}
\big[a(\xi,0))I\big]_{\xi,\eta}^\pi, & (\xi,\eta) \in \widetilde{M}_\lambda^-(QC) \times M_\infty(SO^\diamond),\\
\big[(a(\xi,0)\chi_\lambda^- + a(\xi,1)\chi_\lambda^+)I\big]_{\xi,\eta}^\pi, & (\xi,\eta) \in M_\lambda^0(QC) \times M_\infty(SO^\diamond),\\
\big[a(\xi,1)I\big]_{\xi,\eta}^\pi, & (\xi,\eta) \in \widetilde{M}_\lambda^+(QC) \times M_\infty(SO^\diamond),\\
\big[(a(\xi^+)\chi_- + a(\xi^-)\chi_+)I\big]_{\xi,\eta}^\pi, & (\xi,\eta) \in M_\infty(SO^\diamond) \times M_\tau(QC),\\
\big[\widetilde{a}_\xi I\big]_{\xi,\eta}^\pi, & (\xi,\eta) \in M_\infty(SO^\diamond) \times M_\infty(SO^\diamond)
\end{cases}$$

and

$$\delta_{\xi,\eta}(W^0(b)) := \begin{cases} \big[W^0(b(\eta^+)\chi_- + b(\eta^-)\chi_+)\big]^\pi_{\xi,\eta}, & (\xi,\eta) \in M_\lambda(QC) \times M_\infty(SO^\diamond), \\ \big[b(\eta,0)I\big]^\pi_{\xi,\eta}, & (\xi,\eta) \in M_\infty(SO^\diamond) \times \widetilde{M}_\tau^-(QC), \\ \big[W^0(b(\eta,0)\chi_\tau^- + b(\eta,1)\chi_\tau^+)\big]^\pi_{\xi,\eta}, & (\xi,\eta) \in M_\infty(SO^\diamond) \times M_\tau^0(QC), \\ \big[b(\eta,1)I\big]^\pi_{\xi,\eta}, & (\xi,\eta) \in M_\infty(SO^\diamond) \times \widetilde{M}_\tau^+(QC), \\ \big[W^0(\widetilde{b}_\eta)\big]^\pi_{\xi,\eta}, & (\xi,\eta) \in M_\infty(SO^\diamond) \times M_\infty(SO^\diamond), \end{cases}$$

where $\widetilde{a}_\xi$ *and* $\widetilde{b}_\eta$ *are given by (5.1), extends to a Banach algebra homomorphism* $\delta_{\xi,\eta} : \mathfrak{A}_{\mathbb{R},p,w} \to \mathcal{A}^\pi_{\mathbb{R},\xi,\eta}$*. Moreover,*

$$\sup_{(\xi,\eta)\in\Omega} \|\delta_{\xi,\eta}(A)\|_{\mathcal{A}^\pi_{\mathbb{R},\xi,\eta}} \le \|A^\pi\| := \inf_{K\in\mathcal{K}_{p,w}} \|A+K\| \quad for\ all\ \ A \in \mathfrak{A}_{\mathbb{R},p,w}.$$

Lemma 5.2 *The local algebras* $\mathcal{A}^\pi_{\mathbb{R},\xi,\eta}$ *generated by all the cosets* $\delta_{\xi,\eta}(aW^0(b)) = \big[aW^0(b)\big]^\pi_{\xi,\eta} = \big[aW^0(b)\big]^\pi + J^\pi_{\xi,\eta}$*, where* $a \in PQC_\mathbb{R}$ *and* $b \in PQC_{\mathbb{R},p,w}$*, have the following structure:*

(i) *if* $\lambda \in \mathbb{R}$ *and* $(\xi,\eta) \in M_\lambda^0(QC) \times M_\infty(SO^\diamond)$*, then* $\mathcal{A}^\pi_{\xi,\eta}$ *is generated by the unit* $I^\pi_{\xi,\eta}$ *and two idempotents*

$$P^\pi_{\xi,\eta} := \big[\chi_\lambda^+ I\big]^\pi_{\xi,\eta}, \quad Q^\pi_{\xi,\eta} := \big[W^0(\chi_0^-)\big]^\pi_{\xi,\eta};$$

(ii) *if* $\tau \in \mathbb{R}$ *and* $(\xi,\eta) \in M_\infty(SO^\diamond) \times M_\tau^0(QC)$*, then* $\mathcal{A}^\pi_{\xi,\eta}$ *is generated by the unit* $I^\pi_{\xi,\eta}$ *and two idempotents*

$$P^\pi_{\xi,\eta} := \big[\chi_0^- I\big]^\pi_{\xi,\eta}, \quad Q^\pi_{\xi,\eta} := \big[W^0(\chi_\tau^+)\big]^\pi_{\xi,\eta};$$

(iii) *if* $(\xi,\eta) \in M_\infty(SO^\diamond) \times M_\infty(SO^\diamond)$*, then* $\mathcal{A}^\pi_{\xi,\eta}$ *is generated by the unit* $I^\pi_{\xi,\eta}$ *and the two mutually commuting idempotents*

$$P^\pi_{\xi,\eta} := \big[u_- I\big]^\pi_{\xi,\eta}, \quad Q^\pi_{\xi,\eta} := \big[W^0(u_-)\big]^\pi_{\xi,\eta},$$

where $u_-(x) = (1 - \tanh x)/2$ *for all* $x \in \mathbb{R}$*;*

(iv) *if* $\lambda \in \mathbb{R}$ *and* $(\xi,\eta) \in \widetilde{M}_\lambda^\pm(QC) \times M_\infty(SO^\diamond)$*, then* $\mathcal{A}^\pi_{\xi,\eta}$ *is generated by the unit* $I^\pi_{\xi,\eta}$ *and the one idempotent* $Q^\pi_{\xi,\eta} = \big[W^0(\chi_0^-)\big]^\pi_{\xi,\eta}$*;*

(v) *if* $\tau \in \mathbb{R}$ *and* $(\xi,\eta) \in M_\infty(SO^\diamond) \times \widetilde{M}_\tau^\pm(QC)$*, then* $\mathcal{A}^\pi_{\xi,\eta}$ *is generated by the unit* $I^\pi_{\xi,\eta}$ *and the one idempotent* $P^\pi_{\xi,\eta} = \big[\chi_0^- I\big]^\pi_{\xi,\eta}$*.*

6 Fredholmness of Operators in the Algebra $\mathfrak{A}_{\mathbb{R},p,w}$

Local Spectra of the Cosets $X^\pi_{\xi,\eta}$ Let

$$\Omega_0 = \Big(\bigcup_{\lambda\in\mathbb{R}} M^0_\lambda(QC)\times M_\infty(SO^\diamond)\Big)\cup\Big(M_\infty(SO^\diamond)\times\bigcup_{\tau\in\mathbb{R}} M^0_\tau(QC)\Big). \tag{6.1}$$

For every $\{\xi,\eta) \in \Omega_0$, we define the set

$$\mathfrak{M}_{\xi,\eta} := \big\{\mu \in \mathcal{L}_{p,w,\nu(\xi)} : I^\pi \notin \mathcal{I}^\pi_{\xi,\eta,\mu}\big\}, \tag{6.2}$$

where

$$\mathcal{L}_{p,w,\nu(\xi)} := \big\{(1+\coth[\pi x + \pi i \nu(\xi)])/2 : x \in \overline{\mathbb{R}}\big\}, \tag{6.3}$$

$$\nu(\xi) := \begin{cases} 1/p + (\xi-\lambda)v'(\xi) & if\ \xi \in M^0_\lambda(QC) \text{ for } \lambda\in\mathbb{R}, \\ 1/p + \xi v'(\xi) & if\ \xi \in M_\infty(SO^\diamond). \end{cases} \tag{6.4}$$

Since the cosets $A^\pi_{\xi,\eta}$ for $(\xi,\eta)\in\Omega$ belong to Banach algebras generated by the unit and two idempotents (commuting or not), or by the unit and one idempotent, one can construct the Fredholm symbol calculus for the Banach algebra $\mathfrak{A}_{\mathbb{R},p,w}$ by applying limit operators (see [13]), the two idempotents theorem [3, Theorem 8.7] and Mellin PDO's (see [10]).

Theorem 6.1 *If* $(\xi,\eta)\in\Omega_0$, *then* $\{0,1\}\subset \mathrm{sp}_{\Lambda^\pi_{\xi,\eta}} X^\pi_{\xi,\eta} \subset \mathcal{L}_{p,w,\nu(\xi)}$, *and*

$$\mathrm{sp}_{\Lambda^\pi_{\xi,\eta}} X^\pi_{\xi,\eta} = \mathrm{sp}_{\mathcal{A}^\pi_{\xi,\eta}} X^\pi_{\xi,\eta} = \mathfrak{M}_{\xi,\eta},$$

with $\mathfrak{M}_{\xi,\eta}$, $\mathcal{L}_{p,w,\nu(\xi)}$ *and* $\nu(\xi)$ *given by* (6.2), (6.3) *and* (6.4).

Mappings $\Psi_{\xi,\eta,\mu}$ Along with the set Ω_0 given by (6.1), we define the set

$$\begin{aligned}\widetilde{\Omega} = &\Big(\bigcup_{(\xi,\eta)\in\Omega_0}\{(\xi,\eta)\}\times\mathfrak{M}_{\xi,\eta}\Big)\cup\big(M_\infty(SO^\diamond)\times M_\infty(SO^\diamond)\times\{0,1\}\big)\\ &\cup\Big(\bigcup_{\lambda\in\mathbb{R}}\big((\widetilde{M}^+_\lambda(QC)\cup\widetilde{M}^-_\lambda(QC))\times M_\infty(SO^\diamond)\times\{0,1\}\big)\Big)\\ &\cup\Big(\bigcup_{\lambda\in\mathbb{R}}\big(M_\infty(SO^\diamond)\times(\widetilde{M}^+_\lambda(QC)\cup\widetilde{M}^-_\lambda(QC))\times\{0,1\}\big)\Big).\end{aligned}$$

For each $(\xi,\eta,\mu)\in\widetilde{\Omega}$, we define the mapping

$$\Psi_{\xi,\eta,\mu} : \big\{aI : a \in PQC_{\mathbb{R}}\big\}\cup\big\{W^0(b) : b \in PQC_{\mathbb{R},p,w}\big\} \to \mathbb{C}^{2\times 2}$$

given by the following formulas:

$$\Psi_{\xi,\eta,\mu}(aI) = \begin{bmatrix} a(\xi^+) & 0 \\ 0 & a(\xi^-) \end{bmatrix}, \tag{6.5}$$

$$\Psi_{\xi,\eta,\mu}(W^0(b)) = \begin{bmatrix} b(\eta^+)\mu + b(\eta^-)(1-\mu) & [b(\eta^+) - b(\eta^-)]\varrho(\mu) \\ [b(\eta^+) - b(\eta^-)]\varrho(\mu) & b(\eta^+)(1-\mu) + b(\eta^-)\mu \end{bmatrix},$$

if $(\xi,\eta,\mu) \in \left(\bigcup_{(\xi,\eta)\in\Omega_0}\{(\xi,\eta)\} \times \mathfrak{M}_{\xi,\eta}\right) \cup (M_\infty(SO^\diamond) \times M_\infty(SO^\diamond) \times \{0,1\})$, where $\varrho(\mu)$ is any fixed value of $\sqrt{\mu(1-\mu)}$.

If $\lambda \in \mathbb{R}$, $(\xi,\eta,\mu) \in \widetilde{M}_\lambda^\pm(QC) \times M_\infty(SO^\diamond) \times \{0,1\}$ and $\xi^\pm \in M(PQC)$, then

$$\Psi_{\xi,\eta,\mu}(aI) = \operatorname{diag}\left\{a(\xi^\pm), a(\xi^\pm)\right\}, \tag{6.6}$$
$$\Psi_{\xi,\eta,\mu}(W^0(b)) = \operatorname{diag}\left\{b(\eta^+)\mu + b(\eta^-)(1-\mu),\ b(\eta^+)(1-\mu) + b(\eta^-)\mu\right\}.$$

If $\tau \in \mathbb{R}$, $(\xi,\eta,\mu) \in M_\infty(SO^\diamond) \times \widetilde{M}_\tau^\pm(QC) \times \{0,1\}$ and $\eta^\pm \in M(PQC)$, then

$$\Psi_{\xi,\eta,\mu}(aI) = \operatorname{diag}\left\{a(\xi^+)\mu + a(\xi^-)(1-\mu), a(\xi^+)(1-\mu) + a(\xi^-)\mu\right\},$$
$$\Psi_{\xi,\eta,\mu}(W^0(b)) = \operatorname{diag}\left\{b(\eta^\pm),\ b(\eta^\pm)\right\}. \tag{6.7}$$

Theorem 6.2 *Let $p \in (1,\infty)$ and let $w \in A_p(\mathbb{R})$ be a slowly oscillating weight. The mappings $\Psi_{\xi,\eta,\mu}$ $((\xi,\eta,\mu) \in \widetilde{\Omega})$, given on the generators of the Banach algebra $\mathfrak{A}_{\mathbb{R},p,w}$ by formulas* (6.5)–(6.7)*, extend to Banach algebra homomorphisms $\Psi_{\xi,\eta,\mu} : \mathfrak{A}_{\mathbb{R},p,w} \to \mathbb{C}^{2\times 2}$.*

We assign to every operator $A \in \mathfrak{A}_{\mathbb{R},p,w}$ the bounded matrix function called the *Fredholm symbol* of the operator A by

$$\mathcal{A} : \widetilde{\Omega} \to \mathbb{C}^{2\times 2}, \quad (\xi,\eta,\mu) \mapsto \mathcal{A}(\xi,\eta,\mu) := \Psi_{\xi,\eta,\mu}(A).$$

Let $B(\widetilde{\Omega}, \mathbb{C}^{2\times 2})$ denote the C^*-algebra of all bounded $\mathbb{C}^{2\times 2}$-valued functions on $\widetilde{\Omega}$. The Fredholm symbol mapping

$$\Psi : \mathfrak{A}_{\mathbb{R},p,w} \to B(\widetilde{\Omega}, \mathbb{C}^{2\times 2}), \quad A \mapsto \mathcal{A}(\cdot,\cdot,\cdot),$$

is a Banach algebra homomorphism whose kernel $\operatorname{Ker}\Psi$ contains the ideal $\mathcal{K}_{p,w}$ of all compact operators on $L^p(\mathbb{R}, w)$.

Theorem 6.3 *Let $p \in (1, \infty)$ and let $w \in A_p(\mathbb{R})$ be a slowly oscillating weight. Then the Banach algebra $\mathfrak{A}^{\pi}_{\mathbb{R},p,w}$ is inverse closed in the Calkin algebra $\mathcal{B}^{\pi}_{p,w}$, that is, the spectra of each coset $A^{\pi} \in \mathfrak{A}^{\pi}_{\mathbb{R},p,w}$ in the Banach algebras $\mathfrak{A}^{\pi}_{\mathbb{R},p,w}$ and $\mathcal{B}^{\pi}_{p,w}$ coincide. An operator $A \in \mathfrak{A}_{\mathbb{R},p,w}$ is Fredholm on the space $L^p(\mathbb{R}, w)$ if and only if*

$$\det \mathcal{A}(\xi, \eta, \mu) \neq 0 \ \ for\ all \ \ (\xi, \eta, \mu) \in \widetilde{\Omega}.$$

References

1. M.A. Bastos, A. Bravo, Yu.I. Karlovich, Convolution type operators with symbols generated by slowly oscillating and piecewise continuous matrix functions, in *Oper. Theory Adv. Appl.*, vol. 147 (Birkhäuser, Basel, 2004), pp. 151–174
2. M.A. Bastos, A. Bravo, Yu.I. Karlovich, Symbol calculus and Fredholmness for a Banach algebra of convolution type operators with slowly oscillating and piecewise continuous data. Math. Nachr. **269–270**, 11–38 (2004)
3. A. Böttcher, Yu.I. Karlovich, *Carleson Curves, Muckenhoupt Weights, and Toeplitz Operators*. Progress in Mathematics, vol. 154 (Birkhäuser, Basel, 1997)
4. A. Böttcher, B. Silbermann, *Analysis of Toeplitz Operators*, 2nd edn. (Springer, Berlin, 2006)
5. A. Böttcher, I.M. Spitkovsky, Wiener-Hopf integral operators with PC symbols on spaces with Muckenhoupt weight. Rev. Mat. Iber. **9**, 257–279 (1993)
6. A. Böttcher, I.M. Spitkovsky, Pseudodifferential operators with heavy spectrum. Integral Equations Oper. Theory **19**, 251–269 (1994)
7. R.V. Duduchava, On integral operators of convolution type with discontinuous coefficients. Math. Nachr. **79**, 75–98 (1977)
8. R.V. Duduchava, Integral equations of convolution type with discontinuous coefficients. Soobshch. Akad. Nauk Gruz. SSR **92**, 281–284 (1978) (Russian)
9. R.V. Duduchava, *Integral Equations with Fixed Singularities* (Teubner, Leipzig, 1979)
10. Yu. Karlovich, Algebras of Mellin pseudodifferential operators with quasicontinuous symbols. J. Pseudo-Differ. Oper. Appl. **13**, 18 (2022)
11. Yu.I. Karlovich, I. Loreto Hernández, Algebras of convolution type operators with piecewise slowly oscillating data. I: Local and structural study. Integral Equations Oper. Theory **74**, 377–415 (2012)
12. Yu.I. Karlovich, I. Loreto Hernández, Algebras of convolution type operators with piecewise slowly oscillating data. II: Local spectra and Fredholmness. Integral Equations Oper. Theory **75**, 49–86 (2013)
13. V.S. Rabinovich, S. Roch, B. Silbermann, *Limit Operators and Their Applications in Operator Theory* (Birkhäuser, Basel, 2004)
14. S. Roch, B. Silbermann, Algebras of convolution operators and their image in the Calkin algebra. Report R-Math-05/90. Akademie der Wissenschaften der DDR, Karl-Weierstrass-Institut für Mathematik, Berlin, 1990
15. D. Sarason, Toeplitz operators with piecewise quasicontinuous symbols. Indiana Univ. Math. J. **26**, 817–838 (1977)
16. R. Schneider, Integral equations with piecewise continuous coefficients in L^p-spaces with weight. J. Integral Equations **9**, 135–152 (1985)

Integrability and Convergence of Trigonometric Series and Fourier Transforms

Elijah Liflyand

Abstract A general overview is given for whether a trigonometric series is the Fourier series of an integrable function and when such a series can be approximated in L^1 by its partial sums. This is a necessary preliminary for the attempt to generalize the problem of L^1 convergence of trigonometric series to the non-periodic case, which is the main instance of the present study.

Keywords Trigonometric series · Fourier transform · Bounded variation · Integrability · L^1 convergence

2020 Mathematics Subject Classification Primary 42A38; Secondary 42A16, 42A20, 42A32, 42B35

1 Introduction

In the author's talks at the Tbilisi Analysis & PDE Seminar and related workshop TAPDE2023, a more or less comprehensive picture of what the state of affairs is in the problems of whether a trigonometric series is the Fourier series of an integrable function and L^1 convergence of its partial sums is given, as well as in the parallel study for Fourier transforms. Here we outline this material in a concentrated form.

Before proceeding to the presentation in essence, we mention that notation "$\lesssim$" will be used as abbreviation for "$\le C$", with C being an absolute positive constant.

E. Liflyand (✉)
Department of Mathematics, Bar-Ilan University, Ramat-Gan, Israel

R. Duduchava et al. (eds.), *Tbilisi Analysis and PDE Seminar*,
Research Perspectives Ghent Analysis and PDE Center 7,
https://doi.org/10.1007/978-3-031-62894-8_12

2 Problems in Trigonometric Series

There are two parallel classical topics in harmonic analysis that started simultaneously more than 100 years ago in a work of Young [24].

2.1 Trigonometric Series Versus Fourier Series

One of them is the problem whether a trigonometric series is the Fourier series of an integrable function. More precisely, given a trigonometric series

$$\frac{a_0}{2}+\sum_{k=1}^{\infty}(a_k\cos kx+b_k\sin kx),$$

under what assumptions on its coefficients this series is the Fourier series of an integrable function. Sometimes, it is briefly said "a Fourier series" in place of "the Fourier series of an integrable function". A general approach to this problem, with final answers to many of its aspects, is considered in the recent paper [19].

It made its start from the fact that during a more than one century, there were certain attempts to find necessary and sufficient conditions for the trigonometric series to be a Fourier series: for example, by Salem in [21] or by Edwards in [6], to mention some. However, they were not given in terms of the coefficients of the initial trigonometric series nor were they practical enough. As was shown by Sidon [22], even monotonicity of the coefficients does not guarantee the series to be a Fourier series.

The answer in [19] is given in terms of the Wiener algebra

$$W_0=W_0(\mathbb{R}):=\left\{f(x):\ f(x)=\int_{\mathbb{R}}e^{ixy}g(y)\,dy,\ g\in L_1(\mathbb{R})\right\}.$$

Presenting the trigonometric series even in a more general complex form

$$\sum_{k=-\infty}^{\infty}c_ke^{ikt}, \tag{2.1}$$

we check the belonging to the Wiener algebra of a single, quite simple function ℓ_c, piecewise linear, continuous and such that $\ell_c(k)=c_k$, $k\in\mathbb{Z}$. Among a variety of results in [19], there is the following one.

Criterion *For the series (2.1) to be a Fourier series, it is necessary and sufficient that* $\ell_c\in W_0(\mathbb{R})$.

To be honest, this criterion can be seen in [10]. This was missed by numerous researchers working in the area of trigonometric series. The authors of [19] succeeded to make out this fact at the very last moment. Luckily, this did not destroy [19] since there were many other results of this kind there proven in a more general way as well as many applications were given completely absent in [10].

If the assumptions are posed so that the series converges to an integrable function almost everywhere, then the problem whether this series is a Fourier series is reduced to that of integrability of the sum of the given series, or as it is used to say, to integrability of the series. Here the abuse of terminology is non-essential, the term means integrability of the function which coincides with the sum of the series almost everywhere.

Frequently, the cosine series

$$\frac{a_0}{2} + \sum_{k=1}^{\infty} a_k \cos kx \tag{2.2}$$

and the sine series

$$\sum_{k=1}^{\infty} b_k \sin kx \tag{2.3}$$

are investigated separately, since there is a difference in their behavior. Usually, integrability of the latter requires additional assumptions.

2.2 L^1 Convergence of Partial Sums

When we are within the scope of the integrability of trigonometric series, the second problem naturally arises, whether it is possible to approximate in L^1 the function that the given series represents, or just approximate the series by its partial sums. More precisely, the question is studied what are the assumptions on the coefficients of the series, additional to those for integrability, for the L^1 convergence of the partial sums. Let us give a sample of such a statement. Recall that the N-th partial sums of (2.2) are

$$s_N(f;x)_c = \frac{a_0}{2} + \sum_{k=1}^{N} a_k \cos kx. \tag{2.4}$$

The space of sequences of bounded variation is defined as

$$bv = \left\{ d = \{d_k\} : \|d\|_{bv} = \sum_{k=0}^{\infty} |\Delta d_k| < \infty \right\},$$

where $\Delta d_k = d_k - d_{k+1}$. A typical result can be outlined as follows.

Let the sequence of the coefficients of (2.2) be a bv null sequence. If it satisfies certain assumptions so that (2.2) is the Fourier series of an integrable function f, then

$$\lim_{N\to\infty} \int_0^{\pi} |s_N(f;x)_c - f(x)|\,dx = 0 \tag{2.5}$$

if and only if

$$\lim_{k\to\infty} a_k \ln k = 0. \tag{2.6}$$

We are now in a position to proceed to the non-periodic setting.

3 Non-periodic Analogs

As for integrability of the Fourier transform of a function of bounded variation, the most comprehensive source is [15]. We naturally deal with functions $f \in BV_0(-\infty,\infty) \cap AC(-\infty,\infty)$, which means that f is of bounded variation on $(-\infty,\infty)$, vanishes at infinity and is absolutely continuous on $(-\infty,\infty)$. Instead of the last claim, we can exclude the origin and assume $LAC(\mathbb{R}\setminus\{0\})$, that is, the local absolute continuity on $\mathbb{R}\setminus\{0\}$. Concerning the analogs of the second problem, the first known attempt to extend the known results and methods on the L^1 convergence for trigonometric series to the non-periodic case is [9]. However, that work cannot be indeed considered within the outlined framework. Theorem 1 (4) in the recent paper [20] is better related to our results but a similar assertion is obtained under much more restrictive assumptions. Recently, particular results of the desired type have been obtained in [16].

3.1 Integrability of the Fourier Transform

There is a single necessary and sufficient condition for the integrability of the Fourier transform of functions in the considered class. Following [12], we introduce the space

$$Q = \{g \in L^1(\mathbb{R}) : \int_{\mathbb{R}} \frac{|\widehat{g}(x)|}{|x|}\,dx < \infty\},$$

a subspace of $L^1(\mathbb{R})$, or, more precisely, of $L^1_0(\mathbb{R})$—the space of $L^1(\mathbb{R})$ functions with mean zero. The mentioned criterion is given by

Theorem 3.4 Let f be absolutely continuous on $\mathbb{R}$ and vanishing at infinity. Then $\widehat{f} \in L^1(\mathbb{R})$ if and only if $f' \in Q$.

In fact, the result itself is just integration by parts, while numerous delicate particular conditions, sometimes sophisticated, can be found in [15]. What makes it essential is that Q is an interesting space, wider than all the other spaces of integrability of the Fourier transform of a function of bounded variation. It is worth mentioning that for f with bounded variation, its derivative f' exists almost everywhere and is Lebesgue integrable on $\mathbb{R}$. The Fourier transforms of the considered functions can be understood in the improper sense. Reducing our discussion to a more traditional study of the even and odd cases separately, we are going to deal with functions $f \in BV_0[0,\infty) \cap LAC(0,\infty)$. If Theorem 3.4 is reduced to the half-axis, we have the same results. However, one should realize that, in fact, in each of the two cases Q is reduced to either its subspace Q_e for even functions or the subspace Q_o for odd functions, with $\widehat{f_s}$ or $\widehat{f_c}$ on the half-axis in place of $\widehat{f}$ on $\mathbb{R}$, where

$$\widehat{f_c}(x) = \int_0^\infty f(t) \cos xt\, dt \tag{3.1}$$

is the cosine Fourier transform and

$$\widehat{f_s}(x) = \int_0^\infty f(t) \sin xt\, dt \tag{3.2}$$

is the sine Fourier transform. The right-hand sides of (3.1) and (3.2), with f in question, are natural analogs of trigonometric series (2.2) and (2.3), with bv coefficients, respectively. As mentioned, the recent book [15] gives more or less complete picture of assumptions on $f \in BV_0[0,\infty) \cap LAC(0,\infty)$ under which the Fourier transforms in (3.1) and (3.2) are Lebesgue integrable.

3.2 L^1 Convergence

With the fact in hand that the right-hand sides of (3.1) and (3.2) converge almost everywhere, one may start studying the problems of the L^1 convergence of the partial integrals

$$S_N(f;x)_c = \int_0^N f(t) \cos xt\, dt \tag{3.3}$$

and

$$S_N(f;x)_s = \int_0^N f(t)\sin xt\,dt, \tag{3.4}$$

to $\widehat{f_c}(x)$ and $\widehat{f_s}(x)$, respectively.

Our main result published in [18] reads as follows. Since the approximated Fourier transform is supposed to be integrable, there is no need to approximate it near infinity, where it is already L^1 small. Therefore, the approximation aggregates will be

$$S_N^*(f;x)_c = \begin{cases} \int_0^N f(t)\cos xt\,dt, & 0 \le x \le N, \\ 0, & \text{otherwise,} \end{cases} \tag{3.5}$$

and

$$S_N^*(f;x)_s = \begin{cases} \int_0^N f(t)\sin xt\,dt, & 0 \le x \le N, \\ 0, & \text{otherwise.} \end{cases} \tag{3.6}$$

Theorem 3.5 Let $f \in BV_0[0,\infty) \cap LAC(0,\infty)$. If f' belongs either to Q_o, along with $f(0) = 0$, or to Q_e, then

$$\lim_{N\to\infty} \|\widehat{f_c}(\cdot) - S_N^*(f;\cdot)_c\|_{L^1} = \lim_{N\to\infty} \int_0^N |\widehat{f_c}(x) - S_N(f;x)_c\,|\,dx = 0, \tag{3.7}$$

or

$$\lim_{N\to\infty} \|\widehat{f_s}(\cdot) - S_N^*(f;\cdot)_s\|_{L^1} = \lim_{N\to\infty} \int_0^N |\widehat{f_s}(x) - S_N(f;x)_s\,|\,dx = 0, \tag{3.8}$$

respectively, if and only if

$$\lim_{t\to\infty} |f(t)| \ln t = 0. \tag{3.9}$$

Using the obtained results, we can derive, or, to be more precise, reformulate the results for the general Fourier transform

$$\widehat{f}(x) = \int_{\mathbb{R}} f(t)e^{-ixt}\,dt,$$

while its N-th partial integrals are naturally denoted by

$$S_N(f;x) = \int_{|t|\le N} f(t)e^{-ixt}\,dt.$$

Theorem 3.6 Let $f \in BV_0(-\infty, \infty) \cap AC(-\infty, \infty)$. If $f(0) = 0$ and $f' \in Q$, then

$$\lim_{N\to\infty} \int_{|x|\le N} |S_N(f;x) - \widehat{f}(x)|\,dx = 0$$

if and only if

$$\lim_{|t|\to\infty} |f(t)| \ln |t| = 0. \tag{3.10}$$

It is noteworthy that belonging of the derivative to Q is actively used in the proof.

4 Consequences

We discuss particular but more practical conditions than belonging to Q (or to Q_e and to Q_o) which are involved in the study of the integrability of the Fourier transform of a function of bounded variation.

Recall the scale of subspaces of Q (more precisely, of Q_e and Q_o) proved to be convenient in applications, for which the L^1 convergence results were obtained in [16].

Definition 4.1 For $1 < q < \infty$, the space $O_q := O_q(\mathbb{R}_+)$ is the space of functions g with finite norm

$$\|g\|_{O_q} = \int_0^\infty \left(\frac{1}{x}\int_x^{2x} |g(t)|^q dt\right)^{\frac{1}{q}} dx. \tag{4.1}$$

All these spaces and their sequence analogs first appeared in the paper by D. Borwein [5], but became—for sequences—widely known after the paper by G. A. Fomin [7].

Further, for $q = \infty$, we define the corresponding space as follows.

Definition 4.2 The space O_∞ is the space of functions g with finite norm

$$\|g\|_{O_\infty} = \int_0^\infty \operatorname*{ess\,sup}_{x\le t\le 2x} |g(t)|\,dx. \tag{4.2}$$

The role of the integrable monotone majorant for problems of almost everywhere convergence of singular integrals is known from the work of D.K. Faddeev (see, e.g., [1, Ch. IV, §4]; also [23, Ch. I]); for spectral synthesis problems it was used by Beurling [4], for more details see [2].

In parallel to O_q, we can introduce the spaces E_q as

$$E_q := E_q(\mathbb{R}_+) = \left\{ g : g \in O_q \quad \text{and} \quad \int_0^\infty \frac{1}{x} \left| \int_0^x g(t)\, dt \right| dx < \infty \right\};$$

cf. [15, Ch. 2, Section 2.4] or [17, Ch. 6, §6.4]. Naturally, these spaces are endowed with the norm

$$\|g\|_{E_q} = \|g\|_{O_q} + \int_0^\infty \frac{1}{x} \left| \int_0^x g(t)\, dt \right| dx.$$

If we, as usual, consider the affiliation of f' to one of these spaces, the last integral turns into

$$\int_0^\infty \frac{|f(t)|}{t}\, dt. \tag{4.3}$$

A more general case is that where f' belongs to the real Hardy space $H^1(\mathbb{R})$ (or its even or odd subspaces H_e^1 and H_o^1 naturally considered on $\mathbb{R}_+$). This follows from the well-known Hardy inequality (see, e.g., [8, (7.24)]; in [15] the author suggests to call it the Fourier-Hardy inequality)

$$\int_{\mathbb{R}} \frac{|\widehat{g}(x)|}{|x|}\, dx \lesssim \|g\|_{H^1(\mathbb{R})}. \tag{4.4}$$

There are certain interrelations between these spaces. For example,

$$O_\infty \hookrightarrow O_{p_1} \hookrightarrow O_{p_2} \hookrightarrow H_o^1 \hookrightarrow Q_o \hookrightarrow L^1 \ (p_1 > p_2 > 1), \tag{4.5}$$

which can be found in [13] or [11, Ch. 3]. Similar relations are true in the case of Q_e and its subspaces.

The L^1 convergence result obtained in [16] preceded the one in [18] and reads as follows.

Theorem 4.3 Let $f \in BV_0[0,\infty) \cap LAC(0,\infty)$. If $f' \in O_q(\mathbb{R}_+)$ or $f' \in E_q(\mathbb{R}_+)$ for some $1 < q \le \infty$, then we have (3.7) or (3.8), respectively, if and only if (3.9) holds.

It follows from our main theorem and (4.4) that we can obtain the criterion for more general H_e^1 and H_o^1.

Theorem 4.4 Let $f \in BV_0[0,\infty) \cap LAC(0,\infty)$. If $f' \in H_0^1$, along with $f(0) = 0$, or $f' \in H_e^1$, along with (4.3), then we have (3.7) or (3.8), respectively, if and only if (3.9) holds.

This setting is one of the most powerful in the theory of integrability of Fourier transform as well as its prototype is one of the central results in the theory of trigonometric series. And here the difference between the two settings is revealed. As is shown in [3], the necessary and sufficient condition for the L^1 convergence in the case of discrete Hardy spaces in the periodic setting is not (2.6) but

$$\sum_{k=[n/2]}^{2n} \frac{|a_k| \; or \; |b_k|}{|k-n|+1} = o(1),$$

or corresponding analogs for the sine or general series.

One more chain of subspaces of Q, also wider than the above O_q but incomparable with H^1, proved to be of importance in the considered problems (see [14] or [11]). We start with the definition of these classes.

Definition 4.5 A locally integrable function g defined on $\mathbb{R}_+$ belongs to $A_{1,p}$ if

$$\|g\|_{A_{1,p}} = \sum_{m=-\infty}^{\infty} \left\{ \sum_{j=1}^{\infty} \left[\int_{j2^m}^{(j+1)2^m} |g(t)|\, dt \right]^p \right\}^{\frac{1}{p}} dx < \infty. \qquad (4.6)$$

This space is amalgam in nature, since each of the m-th summands is the norm in the Wiener amalgam space $W(L^1, \ell^p)$ for functions

$$G_m(t) = \begin{cases} 2^m g(2^m t), & if \quad t \ge 1, \\ 0, & \text{otherwise}, \end{cases}$$

where ℓ^p, $1 \le p < \infty$, is a space of sequences $\{d_j\}$ endowed with the norm

$$\|\{d_j\}\|_{\ell^p} = \left(\sum_{j=-\infty}^{\infty} |d_j|^p \right)^{\frac{1}{p}}. \qquad (4.7)$$

The norm of a function $h : \mathbb{R} \to \mathbb{C}$ in the amalgam space $W(L^1, \ell^p)$ is taken as

$$\left\| \left\{ \int_j^{j+1} |h(t)|\, dt \right\} \right\|_{\ell^p}.$$

The proof of the theorem for the Fourier transforms for functions with $f' \in A_{1,2}$ delivers appropriate tools for the study of L^1 convergence in this class.

Theorem 4.6 Let $f \in BV_0[0,\infty) \cap LAC(0,\infty)$. If $f' \in A_{1,2}$, then we have (3.7) and (3.8) provided (4.3) is valid, if and only if (3.9) holds.

We only mention that the greater is p the wider is $A_{1,p}$. Therefore, Theorem 4.6 is more general than such a result with $A_{1,p}$, $1 < p < 2$, in place of $A_{1,2}$.

All the results of this section are just particular cases of our main theorem.

References

1. G. Alexits, *Convergence Problems of Orthogonal Series* (Akadémiai Kiadó, Budapest, 1961)
2. E. Belinsky, E. Liflyand, R. Trigub, The Banach algebra A^* and its properties. J. Fourier Anal. Appl. **3**, 103–129 (1997)
3. A.S. Belov, On conditions of the average convergence (upper boundedness) of trigonometric series. J. Math. Sci. **155**(1), 5–17 (2008)
4. A. Beurling, On the spectral synthesis of bounded functions. Acta Math. **81**, 225–238 (1949)
5. D. Borwein, Linear functionals connected with strong Cesáro summability. J. Lond. Math. Soc. **40**, 628–634 (1965)
6. R.E. Edwards, Criteria for Fourier transforms. J. Austral. Math. Soc. **7**, 239–246 (1967)
7. G.A. Fomin, A class of trigonometric series. Mat. Zametki **23**, 213–222 (1978) (Russian). English transl. in Math. Notes **23**, 117–123 (1978)
8. J. Garcia-Cuerva, J.L. Rubio de Francia, *Weighted Norm Inequalities and Related Topics* (North-Holland, 1985)
9. D.V. Giang, F. Móricz, On the L^1-convergence of Fourier transforms. J. Austral. Math. Soc. (Ser. A) **60**, 495–420 (1996)
10. R.R. Goldberg, Restrictions of Fourier Transforms and Extension of Fourier Sequences. J. Approx. Theory **3**, 149–155 (1970)
11. A. Iosevich, E. Liflyand, *Decay of the Fourier Transform: Analytic and Geometric Aspects* (Birkhäuser, 2014)
12. R.L. Johnson, C.R. Warner, The convolution algebra $H^1(R)$. J. Funct. Spaces Appl. **8**, 167–179 (2010)
13. E. Liflyand, Fourier transforms of functions from certain classes. Anal. Math. **19**, 151–168 (1993)
14. E. Liflyand, Fourier transforms on an amalgam type space. Monatsh. Math. **172**, 345–355 (2013)
15. E. Liflyand, *Functions of Bounded Variation and their Fourier Transforms* (Birkhäuser, 2019)
16. E. Liflyand, L^1 convergence of Fourier transforms. Anal. Math. Phys. **11**(2), Paper No. 91, 11 pp. (2021)
17. E. Liflyand, *Harmonic Analysis on the Real Line: a Path in the Theory* (Birkhäuser, 2021)
18. E. Liflyand, L^1 convergence of Fourier transforms revisited. Results Math. **78**(1), Paper No. 1, 14 pp. (2023)
19. E. Liflyand, R. Trigub, Wiener algebras and trigonometric series in a coordinated fashion. Constr. Approx. **54**, 185–206 (2021)
20. M. Okada, Y. Sawano, Weighted Hankel transform and its applications to Fourier transform. J. Fourier Anal. Appl. **27**, Paper No. 23 (2021)
21. R. Salem, *Essais sur les séries trigonometriques*. Actual. Sci. et Industr., vol. 862 (Hermann, Paris, 1940)
22. S. Sidon, Reihentheoretische Sätze und ihre Anwendungen in der Theorie der Fourierischen Reihen. Math. Zeitschrift **10**, 121–127 (1921)
23. E.M. Stein, G. Weiss, *Introduction to Fourier Analysis on Euclidean Spaces* (Princeton University Press, Princeton, 1971)
24. W.H. Young, On the Fourier series of bounded functions. Proc. Lond. Math. Soc. (2) **12**, 41–70 (1913)

Commutators of Calderón–Zygmund Operators in Grand Variable Exponent Morrey Spaces, and Applications to PDEs

Dali Makharadze, Alexander Meskhi, and Maria Alessandra Ragusa

Abstract In this note we present the boundedness results for commutators of Calderón–Zygmund operators in grand variable exponent Morrey spaces. The operators and spaces are defined on quasi–metric measure spaces with doubling measure. These results are new even for bounded domains in $\mathbb{R}^n$. The proved statements are applied to study regularity properties of solutions of second order partial differential equations with discontinuous coefficients in the frame of grand variable exponent Morrey spaces.

Keywords Grand variable exponent Morrey spaces · Commutators of Calderón–Zygmund operators · Elliptic PDEs · Regularity of solution

2020 AMS Classification 42B20, 42B35, 47B38, 42B37, 47B47

D. Makharadze
Department of Mathematics, Batumi Shota Rustaveli State University, Batumi, Georgia
e-mail: dali.makharadze@bsu.edu.ge

A. Meskhi (✉)
Kutaisi International University, Kutaisi, Georgia

Department of Mathematical Analysis, A. Razmadze Mathematical Institute, I. Javakhishvili Tbilisi State University, Tbilisi, Georgia
e-mail: alexander.meskhi@tsu.ge

M. A. Ragusa
Dipartimento di Matematica e Informatica, Università di Catania, Catania, Italy

Faculty of Fundamental Science, Industrial University, Ho Chi Minh City, Vietnam
e-mail: maragusa@dmi.unict.it

R. Duduchava et al. (eds.), *Tbilisi Analysis and PDE Seminar*,
Research Perspectives Ghent Analysis and PDE Center 7,
https://doi.org/10.1007/978-3-031-62894-8_13

1 Introduction

In this paper the boundedness of commutators of Calderón–Zygmund operators in grand variable exponent Morrey spaces ($GVEMS$s briefly) are studied under the condition that the variable exponent of the space satisfies the log-Hölder continuity condition. As a consequence of the general result we have, in particular, the boundedness of the commutator of the Calderón–Zygmund singular integral operator in these spaces defined on a bounded domain in $\mathbb{R}^n$. Based on the latter result we study the regularity problem for elliptic PDEs on bounded domains in $\mathbb{R}^n$.

One-weight estimates for commutators of sublinear operators in constant exponent weighted grand Morrey spaces ($CEWGMS$s briefly) were investigated in [24] (see [20] for unweighted case). The Boundedness of operators of Harmonic Analysis via the Rubio de Francía's extrapolation in $CEWGMS$s with Muckenhoupt weights was studied in [23] (see also [27] for similar problem beyond the Muckenhoupt classes). The results of [23] were applied by the authors to study regularity of elliptic PDEs in the frame of $CEWGMS$s. We mention also the paper [13] for the boundedness of commutators of sublinear operators in generalized Morrey spaces with constant exponent.

The study of function spaces with variable exponent is a very active area of research nowadays. A variable exponent Lebesgue space ($VELS$ briefly) $L^{p(\cdot)}$ is a special case of the space introduced by W. Orlicz in the 1930s and subsequently generalized by I. Musielak and W. Orlicz. Theses spaces are also called Nakano [32] spaces. For mapping properties of operators of Harmonic Analysis in $VELS$ we refer to the monographs [9, 10, 21], and the survey [15].

Grand Lebesgue spaces $L^{p)}(\Omega)$ defined on bounded domains Ω were introduced in 1992 by T. Iwaniec and C. Sbordone [14]. In subsequent years, quite a number of problems of Harmonic Analysis and the theory of non-linear differential equations were studied in these spaces (see, e.g., the monograph [22] and references cited therein). Grand variable exponent Lebesgue spaces ($GVELS$s briefly) were introduces in [18] (see also [11] for more precise spaces).

Morrey spaces describes regularity problems for solutions of elliptic PDEs more precisely than Lebesgue spaces. Classical Morrey spaces were introduced by C. B. Morrey [31] in 1938 and applied to the regularity problems of solutions of *PDE*s. Variable exponent Morrey spaces appeared in [2], while constant exponent grand Morrey spaces ($CEGMS$s briefly) were introduced and studied in [26] (see also [34] for further generalizations). Grand variable exponent Morrey spaces ($GVEMS$ briefly) were introduced and studied in different viewpoint in the papers [19, 28, 33]. We refer also to the monograph [36] for recent results regarding Morrey-type spaces and applications to PDEs.

2 Preliminaries

Let (X, d, μ) be a quasi-metric measure space ($QMMS$ briefly) with a quasi-metric d and measure μ. A quasi-metric d is a function $d : X \times X \to [0, \infty)$ which satisfies the following conditions: (a) $d(x, y) = 0$ if and only if $x = y$. (b) There is a constant $a_0 > 0$ such that $d(x, y) \le a_0 d(y, x)$ for all $x, y \in X$. (c) There is a constant $a_1 > 0$ such that $d(x, y) \le a_1(d(x, z) + d(z, y))$ for all $x, y, z \in X$.

It is also assumed that all balls $B(x, r) := \{y \in X : d(x, y) < r\}$ in X are measurable, $\mu\{x\} = 0$ for all $x \in X$, μ is a finite measure (i.e. $\mu(X) < \infty$) and that the class of continuous functions with compact supports is dense in the space of integrable functions on X.

We say that a measure μ satisfies doubling condition if there is a positive constant C_{dc} such that for all $x \in X$ and $r > 0$,

$$\mu B(x, 2r) \le C_{dc} \mu B(x, r).$$

We will deal with the $QMMS$ with doubling measure. Such a $QMMS$ is called space of homogeneous type (SHT briefly).

There are several important examples of an SHT: (a) Carleson (regular) curves on $\mathbb{C}$ with arc-length measure $d\nu$ and Euclidean distance on $\mathbb{C}$; (b) Nilpotent Lie groups with Haar measure and homogeneous norm (homogeneous groups); (c) bounded domain Ω in $\mathbb{R}^n$ together with induced Lebesgue measure satisfying so called $\mathcal{A}$ condition, i.e., there is a positive constant C such that for all $x \in \overline{\Omega}$ and $\rho \in (0, d_\Omega)$,

$$\mu(\widetilde{B}(x, \rho)) \ge C\rho^n, \tag{2.1}$$

where d_Ω is a diameter of Ω, and $\widetilde{B}(x, \rho) := \Omega \cap B(x, \rho)$.

2.1 *Grand Variable Exponent Morrey Spaces*

We denote by $P(X)$ the family of all real-valued μ—measurable functions $p(\cdot)$ on X such that

$$1 < p_- \le p_+ < \infty,$$

where

$$p_- := p_-(X) := \inf_X p(x), \quad p_+ := p_+(X) := \sup_X p(x).$$

We say that a function $p(\cdot) \in P(X)$ belongs to the class $\mathcal{P}^{\log}(X)$ if there is a positive constant ℓ such that for all $x, y \in X$ with $d(x, y) \leq 1/2$.

$$|p(x) - p(y)| \leq \frac{\ell}{-\ln(d(x, y))}.$$

A variable exponent Lebesgue space, denoted by $L^{p(\cdot)}(X)$, is the linear space of all Haar-measurable functions f on E for which

$$S_{p(\cdot)}(f) := \int_X |f(x)|^{p(x)} d\mu < \infty.$$

The norm in $L^{p(\cdot)}(X)$ is defined as follows

$$\|f\|_{L^{p(\cdot)}(X)} = \inf\left\{\lambda > 0 : S_{p(\cdot)}(f/\lambda) \leq 1\right\}.$$

Let $\lambda(x)$ be a measurable function on X with values in $[0, 1]$. Denote by $L^{p(\cdot),\lambda(\cdot)}(X)$ the variable exponent Morrey space ($VEMS$ briefly) which is the class of measurable functions on X such that

$$\|f\|_{L^{p(\cdot),\lambda(\cdot)}(X)} = \sup_{x\in X, r\in(0,d_X)} \left\| r^{-\frac{\lambda(x)}{p(\cdot)}} f\chi_{B(x,r)} \right\|_{L^{p(\cdot)}(X)} < \infty,$$

where d_X is the diameter of X.

$GVEMS$s denoted by $L^{p(\cdot),\lambda(\cdot),\theta}(X)$, where $\theta > 0$ and $\lambda(\cdot) \in [0, 1]$, are defined with respect to the following norm:

$$\|f\|_{L^{p(\cdot),\lambda(\cdot),\theta}(X)} = \sup_{0<\varepsilon<p_- -1} \varepsilon^{\frac{\theta}{p_- -\varepsilon}} \|f\|_{L^{p(\cdot)-\varepsilon,\lambda(x)}(X)}.$$

Observe that if $p(\cdot) \in \mathcal{P}^{\log}(X)$, then

$$\|f\|_{L^{p(\cdot),\lambda(\cdot),\theta}(X)} \sim \sup_{0<\varepsilon<p_- -1} \varepsilon^{\frac{\theta}{p_- -\varepsilon}} \left(\sup_{x\in X, 0<r<d_X} \mu B(x, r)^{-\frac{\lambda(x)}{p(x)-\varepsilon}} \|f\|_{L^{p(x)-\varepsilon}(B(x,r))} \right).$$

We are interested in the case when λ is constant. In particular, grand variable exponent Morrey space $L^{p(\cdot),\lambda,\theta}(X)$ defined by the norm:

$$\|f\|_{L^{p(\cdot),\lambda,\theta}(X)} := \sup_{0<\varepsilon<p_- -1} \sup_B \frac{\varepsilon^{\frac{\theta}{p_- -\varepsilon}}}{(\mu B)^{\lambda}} \|f\|_{L^{p(\cdot)-\varepsilon}(B)},$$

where $p(\cdot) \in P(X), 0 < \lambda < 1, \theta > 0$.

2.2 Commutators of Calderón–Zygmund Operators

The space of functions of bounded mean oscillation, denoted by $\mathrm{BMO}(X,\mu)$, is the set of all real-valued locally integrable functions such that

$$\|f\|_{\mathrm{BMO}(X,\mu)} = \sup_{\substack{x\in X\\ 0<r<d_X}} \frac{1}{\mu B(x,r)} \int_{B(x,r)} \left|f(y) - f_{B(x,r)}\right| \mathrm{d}\mu(y) < \infty,$$

where $f_{B(x,r)}$ is the integral average over the ball $B(x,r)\cdot \mathrm{BMO}(X,\mu)$ is a Banach space with respect to the norm $\|\cdot\|_{\mathrm{BMO}(X,\mu)}$ when we regard the space $\mathrm{BMO}(X,\mu)$ as the class of equivalent functions modulo additive constants.

An equivalent norm for $\|\cdot\|_{\mathrm{BMO}(X,\mu)}$ is defined as

$$\|f\|_{\mathrm{BMO}(X,\mu)} \sim \sup_{\substack{x\in X\\ 0<r<d_X}} \inf_{c\in\mathbb{R}} \frac{1}{\mu B(x,r)} \int_{B(x,r)} |f(y) - c| \mathrm{d}\mu(y).$$

Let U be an operator and b a locally integrable function. We define the commutator $U_b f$ as

$$U_b f = bU(f) - U(bf).$$

We are interested in commutators of Calderón–Zygmund operators defined on an SHT (see [3] for the boundedness results of such commutators in weighted classical Lebesgue spaces with Muckenhoupt weights).

Let K be the Calderón–Zygmund operator (see, e.g., [17])

$$Kf(x) = p.v. \int_X k(x,y) f(y) d\mu(y),$$

where k is the Calderón-Zygmund kernel $k : X \times X\backslash\{(x,x) : x \in X\} \to \mathbf{R}$ satisfying the conditions:

(i) $|k(x,y)| \le \frac{C}{\mu B(x,d(x,y))}, \quad x,y\in X, \quad x\ne y;$

(ii)

$$|k(x_1,y)-k(x_2,y)|+|k(y,x_1)-k(y,x_2)| \le Cw\left(\frac{d(x_2,x_1)}{d(x_2,y)}\right) \frac{1}{\mu B(x_2,d(x_2,y))}$$

for all x_1, x_2 and y with $d(x_2,y) \ge Cd(x_1,x_2)$, where w is a positive non-decreasing function on $(0,\infty)$ which satisfies the Δ_2 condition $w(2t) \le cw(t)$ $(t>0)$ and the Dini condition $\int_0^1 w(t)/t \, \mathrm{d}t < \infty$.

It is also assumed that k is such that Kf exists almost everywhere on X in the principal value sense for all $f \in L^2(X)$ and that K is bounded in $L^2(X)$.

3 Boundedness Result

By the symbol $D(X)$ we denote the class of bounded functions on X with compact supports.

Now we formulate the main boundedness results of this note:

Theorem 3.1 *Let $p(\cdot) \in \mathcal{P}^{\log}(X)$ and the θ and λ be constants such that $\theta > 0, 0 < \lambda < 1/p_+$. Let $b \in \mathrm{BMO}(X, \mu)$. Then there is a positive constant C independent of f such that*

$$\|K_b f\|_{L^{p(\cdot),\lambda,\theta}(X)} \leqq C\|f\|_{L^{p(\cdot),\lambda,\theta}(X)}, \quad f \in D(X).$$

Corollary 3.2 *Let Ω be a bounded domain in $\mathbb{R}^n$ such that condition* (2.1) *is satisfied. Let K_b be the commutator of the Calderón–Zygmund operator defined on Ω. Suppose that $p(\cdot) \in \mathcal{P}^{\log}(\Omega)$ and the θ and λ be constants such that $\theta > 0, 0 < \lambda < 1/p_+$. Let $b \in \mathrm{BMO}(\Omega, dx)$. Then there is a positive constant C independent of f such that*

$$\|K_b f\|_{L^{p(\cdot),\lambda,\theta}(\Omega)} \leqq C\|f\|_{L^{p(\cdot),\lambda,\theta}(\Omega)}, \quad f \in D(\Omega).$$

4 Applications to PDEs

In the last thirty years a number of papers have been devoted to the study of local and global regularity properties of strong solutions to elliptic equations with discontinuous coefficients. To be more precise, let us consider the second order equation

$$\mathcal{L}u \equiv \sum_{i,j=1}^{n} a_{ij}(x) D_{x_i x_j} u = f(x) \quad \text{for almost all } x \in \Omega, \tag{4.1}$$

where $\mathcal{L}$ is a uniformly elliptic operator over the bounded domain $\Omega \subset \mathbb{R}^n$, $n \geq 2$.

We assume that a domain Ω satisfies $\mathcal{A}$ condition (see (2.1)). In this case Ω with induced Lebesgue measure and Euclidean metrics is an SHT. Hence, the previous statements are valid for such domains.

Regularizing properties of $\mathcal{L}$ in Hölder spaces (i.e. $\mathcal{L}u \in C^{\alpha}(\bar{\Omega})$ implies $u \in C^{2+\alpha}(\bar{\Omega})$) have been well studied in the case of Hölder continuous coefficients $a_{ij}(x)$. Also, unique classical solvability of the Dirichlet problem for (4.1) has been derived in this case (we refer to [12] and the references therein). In the case of uniformly continuous coefficients a_{ij}, an L^p-Schauder theory has been elaborated for the operator $\mathcal{L}$ (see [1, 12]). In particular, $\mathcal{L}u \in L^p(\Omega)$ always implies that the strong solution to (4.1) belongs to the Sobolev space $W^{2,p}(\Omega)$ for each $p \in (1, \infty)$. However, the situation becomes rather difficult if one tries to allow discontinuity at the principal coefficients of $\mathcal{L}$. In general, it is well-known (cf. [29]) that arbitrary discontinuity of a_{ij} implies that the L^p-theory of $\mathcal{L}$ and the strong solvability of the Dirichlet problem for (4.1) break down. A notable exception of that rule is the two-dimensional case ($\Omega \subset \mathbb{R}^2$). It was shown by G. Talenti that the solely condition on measurability and boundedness of the a_{ij}'s ensures isomorphic properties of $\mathcal{L}$ considered as a mapping from $W^{2,2}(\Omega) \cap W_0^{1,2}(\Omega)$ into $L^2(\Omega)$. To handle with the multidimensional case ($n \geq 3$) requires that additional properties on $a_{ij}(x)$ should be added to the uniform ellipticity in order to guarantee that $\mathcal{L}$ possesses the regularizing property in Sobolev functional scales. In particular, if $a_{ij}(x) \in W^{1,n}(\Omega)$ (cf. [30]), or if the difference between the largest and the smallest eigenvalues of $\{a_{ij}(x)\}$ is small enough (the Cordes condition), then $\mathcal{L}u \in L^2(\Omega)$ yields that $u \in W^{2,2}(\Omega)$ and these results can be extended to $W^{2,p}(\Omega)$ for $p \in (2-\varepsilon, 2+\varepsilon)$ with sufficiently small ε.

Later the Sarason class VMO of functions with vanishing mean oscillation was used in the study of local and global Sobolev regularity of the strong solutions to (4.1).

Next, we define the space *BMO* and then the smallest *VMO* class, where we consider coefficients a_{ij} and later that one where we consider the known term f.

In the sequel let Ω be an open bounded set in $\mathbb{R}^n$.

Definition 4.1 Let $f \in L^1_{\text{loc}}(\Omega)$. We define the integral mean $f_{x,R}$ by

$$f_{x,R} := \frac{1}{|\Omega \cap B(x,R)|} \int\limits_{\Omega \cap B(x,R)} f(y)dy,$$

where $B(x,R)$ ranges in the class of balls centered in x with radius R and $|\Omega \cap B(x,R)|$ is the Lebesgue measure of $\Omega \cap B(x,R)$. If we are not interested in specifying which the center is, we just use the notation f_R.

We now give the definition of Bounded Mean Oscillation functions (BMO) that appeared at first in the note by F. John and L. Nirenberg [16].

Definition 4.2 Let $f \in L^1_{\text{loc}}(\Omega)$. We say that f belongs to $BMO(\Omega)$ if the seminorm $\|f\|_*$ is finite, where

$$\|f\|_* := \sup_{B(x,R)} \frac{1}{|B(x,R)|} \int\limits_{B(x,R)} |f(y) - f_{x,R}|dy.$$

Next, we consider the definition of the space of Vanishing Mean Oscillation functions, given at first by D. Sarason [35].

Definition 4.3 Let $f \in BMO(\Omega)$ and

$$\eta(f, R) := \sup_{\rho \leq R} \frac{1}{|B_\rho|} \int_{B_\rho} |f(y) - f_\rho| dy,$$

where $B\rho$ ranges over the class of the balls of $\mathbb{R}^n$ of radius ρ. Further, a function $f \in VMO(\Omega)$ if $\lim_{R\to 0} \eta(f, R) = 0$.

The Sarason class is then expressed as the subspace of the functions in the John-Nirenberg class whose BMO norm over a ball vanishes as the radius of the balls tends to zero. This property implies a number of good features of VMO functions not shared by general BMO functions; in particular, they can be approximated by smooth functions.

This class of functions was considered by many others. At first, we recall the paper by F. Chiarenza et al. [7], where the authors answer a question raised thirty years before by C. Miranda [30]. In his paper he considers a linear elliptic equation where the coefficients a_{ij} of the higher order derivatives are in the class $W^{1,n}(\Omega)$ and asks whether the gradient of the solution is bounded, if $p > n$. In [7] the authors suppose that $a_{ij} \in VMO$ and prove that Du is Hölder continuous for all $p \in]1, +\infty[$.

Also, it is possible to check that bounded uniformly functions are in VMO as well as functions of fractional Sobolev spaces $W^{\theta, \frac{n}{\theta}}$, $\theta \in]0, 1[$.

The study of Sobolev regularity of strong solutions of (4.1) was initiated in 1991 with the pioneering work by F. Chiarenza et al. [6]. It was obtained that, if $a_{ij}(x) \in VMO \cap L^\infty(\Omega)$ and $\mathcal{L}u \in L^p(\Omega)$, then $u \in W^{2,p}(\Omega)$ for *each* value of p in the range $(1, \infty)$. Moreover, well-posedness of the Dirichlet problem for (4.1) in $W^{2,p}(\Omega) \cap W_0^{1,p}(\Omega)$ was proved. As a consequence, Hölder continuity of the strong solution or of its gradient follows if the exponent p is sufficiently large.

Thanks to the fundamental accessibility of these two papers [7, 8] and many other authors have used VMO class to obtain regularity results for PDEs and systems with discontinuous coefficients.

Continuing the study of regularity of PDEs, we see that Hölder continuity can be inferred for small p if one has more information on $\mathcal{L}u$, such as its belonging to suitable Morrey class $L^{p,\lambda}(\Omega)$.

Using these spaces a natural problem arises namely to study the regularizing properties of the operator $\mathcal{L}$ in Morrey spaces in the case of VMO principal coefficients. In [4] proved that each $W^{2,p}$-viscosity solution to (4.1) lies in $C^{1+\alpha}(\Omega)$ if $f(x)$ belongs to the Morrey space $M^{n,n\alpha}(\Omega)$ with $\alpha \in (0, 1)$.

One of the main results of this note is to obtain local regularity, in grand Morrey spaces, for highest order derivatives of solutions of elliptic non-divergence form with coefficients, which can be discontinuous.

We recall that, in the case of continuous coefficients of the above kind of equation, the results obtained by S. Agmon, A. Douglis and L. Nirenberg [1]. Later, discontinuous coefficients were considered by S. Campanato [5].

Then, this paper can be regarded as a continuation of the study of L^p regularity of solutions of second order elliptic PDEs to the maximum order derivatives of the solutions to a certain class of linear elliptic equations in nondivergence form with discontinuous coefficients (see also [25] for related topics).

Let us consider the second order differential operator

$$\mathcal{L} \equiv a_{ij}(x)D_{ij}, \qquad D_{ij} \equiv \frac{\partial^2}{\partial x_i \partial x_j}.$$

Here we have adopted the usual summation convention on repeated indices.

In the sequel, we need the following regularity and ellipticity assumptions on the coefficients of $\mathcal{L}$, $\forall i, j = 1, \dots, n$:

$$\begin{cases} a_{ij}(x) \in L^\infty(\Omega) \cap VMO, \\ a_{ij}(x) = a_{ji}(x), \quad \text{a.a. } x \in \Omega, \\ \exists\, \kappa > 0 : \ \kappa^{-1}|\xi|^2 \le a_{ij}(x)\xi_i\xi_j \le \kappa|\xi|^2, \quad \forall \xi \in \mathbb{R}^n, \text{ a.a. } x \in \Omega. \end{cases} \tag{4.2}$$

Set η_{ij} for the VMO-modulus of the function $a_{ij}(x)$ and let $\eta(r) = \left(\sum_{i,j=1}^{n} \eta_{ij}^2\right)^{1/2}$. We denote by $\Gamma(x,t)$ the normalized fundamental solution of $\mathcal{L}$, i.e.,

$$\Gamma(x,\xi) = \frac{1}{n(2-n)\omega_n\sqrt{\det\{a_{ij}(x)\}}}\left(\sum_{i,j=1}^{n} A_{ij}(x)\xi_i\xi_j\right)^{(2-n)/2}$$

for a.a. x and all $\xi \in \mathbb{R}^n \setminus \{0\}$, where $A_{ij}(x)$ stand for the entries of the inverse matrix of the matrix $\{a_{ij}(x)\}_{i,j=1,\dots,n}$, and ω_n is the measure of the unit ball in $\mathbb{R}^n$. We set also

$$\Gamma_i(x,\xi) = \frac{\partial}{\partial \xi_i}\Gamma(x,\xi), \quad \Gamma_{ij}(x,\xi) = \frac{\partial}{\partial \xi_i \partial \xi_j}\Gamma(x,\xi),$$

$$M = \max_{i,j=1,\dots,n}\ \max_{|\alpha|\le 2n} \left\| \frac{\partial^\alpha \Gamma_{ij}(x,\xi)}{\partial \xi^\alpha} \right\|_{L^\infty(\Omega\times\Sigma)}.$$

It is well known that $\Gamma_{ij}(x,\xi)$ are Calderón–Zygmund kernels in the ξ variable.

Theorem 4.4 *Let the coefficients of $\mathcal{L}$ satisfies* (4.2) *and $p(\cdot) \in \mathcal{P}^{\log}(\Omega)$, $0 < \lambda < 1/p_+$, $\theta > 0$. Let Ω be a domain satisfying $\mathcal{A}$ condition (see* (2.1)*). Then there exist positive constants $c = c(n, \kappa, p(\cdot), \lambda, \theta, M)$ and $\rho_0 = \rho_0(c, n)$ such that for every ball $B_\rho \subset\subset \Omega$, $\rho < \rho_0$ and every $u \in W_0^{2,p}(B_\rho)$ such that $D_{ij}u \in L^{p(\cdot),\lambda,\theta}(B_\rho)$, we have*

$$\|D_{ij}u\|_{L^{p(\cdot),\lambda,\theta}(B_\rho)} \le c\|\mathcal{L}u\|_{L^{p(\cdot),\lambda,\theta}(B_\rho)}, \quad \forall i, j = 1, \dots, n. \tag{4.3}$$

Acknowledgments The first and second authors were supported by Batumi Shota Rustaveli State University Grant (Resolution # 06- 01/04, 31.01.2023). The third author wish to thank Faculty of Fundamental Science, Industrial University of Ho Chi Minh City, Vietnam, for the opportunity to work in it.

References

1. S. Agmon, A. Douglis, L. Nirenberg, Estimates near the boundary for solutions of elliptic partial differential equations satisfying general boundary conditions. I. Comm. Pure Appl. Math. **12**, 623–727 (1959)
2. A. Almeida, J. Hasanov, S. Samko, Maximal and potential operators in variable exponent Morrey spaces. Georgian Math. J. **15**(2), 195–208 (2008)
3. A. Bernardis, S. Hartzstein, G. Pradolini, Weighted inequalities for commutators of fractional integrals on spaces of homogeneous type. J. Math. Anal. Appl. **322**, 825–846 (2006)
4. L. Caffarelli, Elliptic second order equations. Rend. Sem. Mat. Fis. Milano **58**, 253–284 (1988)
5. S. Campanato, Sistemi parabolici del secondo ordine, non variazionali, a coefficienti discontinui. (Italian) Ann. Univ. Ferrara Sez. VII (N.S.) **23**, 169–187 (1977)
6. F. Chiarenza, M. Frasca, P. Longo, Interior $W^{2,p}$ estimates for nondivergence elliptic equations with discontinuous coefficients. Ricerche Mat. **40**(1), 149–168 (1991)
7. F. Chiarenza, M. Frasca, P. Longo, $W^{2,p}$-solvability of the Dirichlet problem for nondivergence elliptic equations with VMO coefficients. Trans. Amer. Math. Soc. **336**(2), 841–853 (1993)
8. F. Chiarenza, M. Franciosi, M. Frasca, L^p-estimates for linear elliptic systems with discontinuous coefficients. Atti Accad. Naz. Lincei Cl. Sci. Fis. Mat. Natur. Rend. Lincei (9) Mat. Appl. **5**(1), 27–32 (1994)
9. D.V. Cruz-Uribe, A. Fiorenza, *Variable Lebesgue Spaces, Foundations and Harmonic Analysis*. Applied and Numerical Harmonic Analysis (Birkhäuser/Springer, Heidelberg, 2013)
10. L. Diening, P. Harjulehto, P. Hästö, M. Ružička, *Lebesgue and Sobolev Spaces with Variable Exponents*. Lecture Notes in Mathematics (Springer, Heidelberg, 2011)
11. D.E. Edmunds, V. Kokilashvili, A. Meskhi, Sobolev-type inequalities for potentials in grand variable exponent Lebesgue spaces. Math. Nachr. **292**(10), 2174–2188 (2019)
12. D. Gilbarg, N.S. Trudinger, *Elliptic Partial Differential Equations of Second Order*, 2nd edn., vol. 224 Grundlehren der mathematischen Wissenschaften [Fundamental Principles of Mathematical Sciences] (Springer, Berlin, 1983)
13. V.S. Guliyev, S.S. Aliyev, T. Karaman, P.S. Shukurov, Boundedness of sublinear operators and commutators on generalized Morrey spaces. Integral Equations Oper. Theory **71**, 327–355 (2011)
14. T. Iwaniec, C. Sbordone, On the integrability of the Jacobian under minimal hypotheses. Arch. Rational Mech. Anal. **119**, 129–143 (1992)
15. M. Izuki, E. Nakai, Y. Sawano, Function spaces with variable exponent, An Introduction. Sci. Math. Jpn. **77**(2), 187–315 (2014)
16. F. John, L. Nirenberg, On functions of bounded mean oscillation. Comm. Pure Appl. Math. **14**, 415–426 (1961)
17. V. Kokilashvili, A. Meskhi, Boundedness of maximal and singular operators in Morrey spaces with variable exponent. Armen. J. Math. **1**(1), 18–28 (2008)
18. V. Kokilashvili, A. Meskhi, Maximal and Calderón-Zygmund operators in grand variable exponent Lebesgue spaces. Georgian Math. J. **21**(4), 447–461 (2014)
19. V. Kokilashvili, A. Meskhi, Boundedness of operators of Harmonic Analysis in grand variable exponent Morrey spaces. Mediterr. J. Math. **71** (2023). https://doi.org/10.1007/s00009-023-02267-8

20. V. Kokilashvili, A. Meskhi, H. Rafeiro, Boundedness of commutators of singular and potential operators in generalized grand Morrey spaces and some applications. Studia Math. **217**(2), 159–178 (2013)
21. V. Kokilashvili, A. Meskhi, H. Rafeiro, S. Samko, *Integral Operators in Non-standard Function Spaces: Variable Exponent Lebesgue and Amalgam Spaces*, vol. 1 (Birkäuser/Springer, Heidelberg, 2016)
22. V. Kokilashvili, A. Meskhi, H. Rafeiro, S. Samko, *Integral Operators in Non-standard Function Spaces. Vol. 2. Variable Exponent Hölder, Morrey-Campanato and Grand Spaces*. Operator Theory: Advances and Applications, vol. 249 (Birkhäuser/Springer, 2016)
23. V. Kokilashvili, A. Meskhi, M.A. Ragusa, Weighted extrapolation in grand Morrey spaces and applications to partial differential equations. Rend. Lincei Mat. Appl. **30**, 67–92 (2019)
24. V. Kokilashvili, A. Meskhi, H. Rafeiro, Commutators of sublinear operators in grand Morrey spaces. Studia Sci. Math. Hungar. **56**(2), 211–232 (2019)
25. V. Kokilashvili, A. Meskhi, M.A. Ragusa, Weighted extrapolation in grand Morrey spaces and applications to partial differential equations. Atti Accad. Naz. Lincei Rend. Lincei Mat. Appl. **30**(1), 67–92 (2019)
26. A. Meskhi, Maximal functions, potentials and singular integrals in grand Morrey spaces. Complex Var. Elliptic Equations **56**(10–11), 1003–1019 (2011)
27. A. Meskhi, Extrapolation in new weighted grand Morrey spaces beyond the Muckenhoupt classes. J. Math. Anal. Appl. (2023). https://doi.org/10.1016/j.jmaa.2023.127181
28. A. Meskhi, Y. Sawano, Density, duality and preduality in grand variable exponent Lebesgue and Morrey spaces. Mediterranean J. Math. **15** (2018). https://doi.org/10.1007/s00009-018-1145-5
29. N. Meyers, An L^pe-estimate for the gradient of solutions of second order elliptic divergence equations. Ann. Scuola Norm. Sup. Pisa Cl. Sci. (3) **17**, 189–206 (1963)
30. C. Miranda, Sulle equazioni ellittiche del secondo ordine di tipo non variazionale, a coefficienti discontinui (Italian). Ann. Mat. Pura Appl. **63**, 353–386 (1963)
31. C.B. Morrey, On the solutions of quasi-linear elliptic partial differential equations. Trans. Amer. Math. Soc. **43**(1), 126–166 (1938)
32. H. Nakano, *Topology of Linear Topological Spaces* (Maruzen, Tokyo, 1951)
33. T. Ohno, T. Shimomura, Sobolev embeddings for Riesz potentials of functions in grand Morrey spaces of variable exponents over non-doubling measure spaces. Czech. Math. J. **64**(1), 209–228 (2014)
34. H. Rafeiro, A note on boundedness of operators in grand grand Morrey spaces, in *Advances in Harmonic Analysis and Operator Theory*. Oper. Theory Adv. Appl., vol. 229 (Birkhäuser/Springer Basel AG, Basel, 2013), pp. 349–356
35. D. Sarason, Functions of vanishing mean oscillation. Trans. Amer. Math. Soc. **207**, 391–405 (1975)
36. Y. Sawano, G. Di Fazio, D.I. Hakim, Morrey spaces, in *Introduction and Applications to Integral Operators and PDE's*, vols. I and II (CRC Press, Taylor and Francis, 2020)

Symmetric Stein–Tomas, and Why Do We Care?

Rainer Mandel and Diogo Oliveira e Silva

Abstract We summarize our recent joint work on the Stein–Tomas inequality under symmetries, and discuss three applications.

Keywords Biharmonic NLS · Ground states · Helmholtz resolvent · Orthogonal group · Sharp restriction theory · Stein-Tomas inequality · Symmetry breaking

2000 Mathematics Subject Classification 42B10

1 Introduction

The restriction problem for the Fourier transform asks the following: Given $1 \leq p \leq 2$, for which $1 \leq q \leq \infty$ does the inequality

$$\int_{\mathbb{S}^{d-1}} |\widehat{f}(\omega)|^q \,\mathrm{d}\sigma(\omega) \lesssim \|f\|^q_{L^p(\mathbb{R}^d)} \tag{1}$$

hold? Here, the unit sphere[1] $\mathbb{S}^{d-1} = \{x \in \mathbb{R}^d : |x| = 1\}$ comes equipped with the usual surface measure σ. It is easy to check that (1) holds if $p = 1$ (for every q)

[1] What is crucial about the sphere is that it is *curved*: analogous restriction-type estimates hold for any smooth compact hypersurface of nonvanishing gaussian curvature.

R. Mandel
Karlsruhe Institute of Technology, Institute for Analysis, Karlsruhe, Germany
e-mail: Rainer.Mandel@gmx.de

D. Oliveira e Silva (✉)
Center for Mathematical Analysis, Geometry and Dynamical Systems & Departamento de Matemática, Instituto Superior Técnico, Lisboa, Portugal
e-mail: diogo.oliveira.e.silva@tecnico.ulisboa.pt

R. Duduchava et al. (eds.), *Tbilisi Analysis and PDE Seminar*,
Research Perspectives Ghent Analysis and PDE Center 7,
https://doi.org/10.1007/978-3-031-62894-8_14

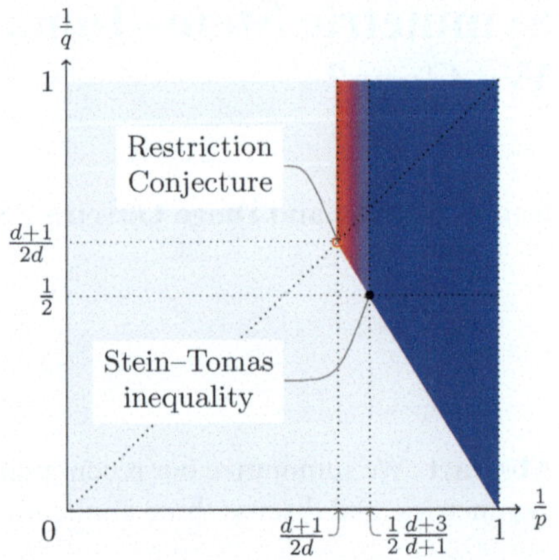

Fig. 1 Riesz diagram for the restriction operator to $\mathbb{S}^{d-1}$

and does not hold if $p = 2$ (for any q). Stein's restriction conjecture predicts that inequality (1) holds if and only if

$$1 \leq p < \frac{2d}{d+1} \text{ and } q \leq \frac{d-1}{d+1}p', \tag{2}$$

where $p' = p/(p-1)$ is the exponent dual to p; see Fig. 1. To verify the necessity of these conditions, one tests the dual inequality to (1) against the constant function on $\mathbb{S}^{d-1}$ and the so-called Knapp example; see Sects. 1.1 and 1.2 below. The sufficiency of conditions (2) for inequality (1) to hold is open in dimensions[2] $d \geq 3$.

If $q = 2$, then (1) holds if and only if $1 \leq p \leq 2\frac{d+1}{d+3}$. This is the content of the classical Stein–Tomas inequality [11, 12], which plays a central role in this note.

1.1 The Radial Case

The restriction problem for radial functions has a simple solution that had already been observed by L. Schwartz; see [9, §11, Notes]. Indeed, if $f \in L^p(\mathbb{R}^d)$ is radial and $1 \leq p < \frac{2d}{d+1}$, then $\widehat{f}$ is continuous on $\mathbb{R}^d \setminus \{0\}$. The proof of this statement is based on the integral representation

$$\widehat{f}(\rho) \simeq_d \rho^{\frac{2-d}{2}} \int_0^\infty f(r) J_{\frac{d-2}{2}}(\rho r) r^{\frac{d}{2}} \, \mathrm{d}r. \tag{3}$$

[2] The restriction conjecture to the circle $\mathbb{S}^1 \subset \mathbb{R}^2$ can be verified via a successive application of the inequalities of Hausdorff–Young and Hardy–Littlewood–Sobolev; see [9, §11.4].

From the Bessel decay $|J_{\frac{d-2}{2}}(x)| \lesssim_d \min\{|x|^{\frac{d-2}{2}}, |x|^{-\frac{1}{2}}\}$ and Hölder's inequality, one infers for $|\rho| \approx 1$ that

$$\begin{aligned}
|\widehat{f}(\rho)| &\lesssim_{p,d} \int_0^1 |f(r)|r^{d-1}\,\mathrm{d}r + \int_1^\infty |f(r)|r^{\frac{d-1}{2}}\,\mathrm{d}r \\
&\lesssim_d \int_0^\infty |f(r)|r^{\frac{d-1}{p}} \cdot (1+r)^{\frac{d-1}{2}-\frac{d-1}{p}}\,\mathrm{d}r \\
&\leq \left(\int_0^\infty |f(r)|^p r^{d-1}\,\mathrm{d}r\right)^{\frac{1}{p}} \left(\int_0^\infty (1+r)^{(\frac{d-1}{2}-\frac{d-1}{p})p'}\,\mathrm{d}r\right)^{\frac{1}{p'}}.
\end{aligned}$$

In the last line, the first integral equals $\|f\|_p^p$ while the second integral is finite if and only if $p < \frac{2d}{d+1}$, which is the first condition in (2).

1.2 *Knapp's Example on* $\mathbb{S}^{d-1}$

Let f be the indicator function of the cap $\mathcal{C}_\delta \subset \mathbb{S}^{d-1}$,

$$\mathcal{C}_\delta = \{\omega \in \mathbb{S}^{d-1} : |\omega - e_d| \leq \delta\},$$

where $e_d = (0, \ldots, 0, 1) \in \mathbb{R}^d$. In particular, $\|f\|_{L^{q'}(\mathbb{S}^{d-1})} \simeq \delta^{\frac{d-1}{q'}}$. In view of the uncertainty principle for the Fourier transform, the adjoint restriction, or *extension*, operator,

$$\widehat{f\sigma}(x) = \int_{\mathbb{S}^{d-1}} f(\omega)e^{ix\cdot\omega}\,\mathrm{d}\sigma(\omega), \tag{4}$$

satisfies the pointwise lower bound $|\widehat{f\sigma}(x)| \gtrsim \delta^{d-1}$ as long as $x \in \mathcal{R}_\delta^\star$. Here, $\mathcal{R}_\delta^\star \subset \mathbb{R}^d$ stands for the box *dual* to the smallest box in $\mathbb{R}^d$ containing the cap $\mathcal{C}_\delta$, of length equal to $C^{-1}\delta^{-2}$ (in the direction of e_d) and cross sectional radius equal to $C^{-1}\delta^{-1}$, for some sufficiently large constant $C < \infty$. Thus

$$\|\widehat{f\sigma}\|_{p'} \gtrsim \delta^{d-1}\|\mathbf{1}_{\mathcal{R}_\delta^\star}\|_{p'} \simeq \delta^{d-1}\left(\delta^{-(d-1)}\delta^{-2}\right)^{\frac{1}{p'}} = \delta^{d-1-\frac{d+1}{p'}}.$$

By letting $\delta \to 0^+$, the $L^{q'}(\mathbb{S}^{d-1}) \to L^{p'}$ extension estimate

$$\delta^{d-1-\frac{d+1}{p'}} \lesssim \|\widehat{f\sigma}\|_{p'} \lesssim \|f\|_{L^{q'}(\mathbb{S}^{d-1})} \simeq \delta^{\frac{d-1}{q'}}$$

can only hold if $d-1-\frac{d+1}{p'} \geq \frac{d-1}{q'}$, or $q \leq \frac{d-1}{d+1}p'$, which is the second condition in (2).

2 Symmetric Setup

Given a subgroup G of the orthogonal group $O(d)$, a function $f : \mathbb{R}^d \to \mathbb{C}$ is said to be *G-symmetric in* $\mathbb{R}^d$ if $f \circ A = f$ for every $A \in G$. We shall focus on the case of the subgroup $G_k := O(d-k) \times O(k)$, for some $0 \le k \le d$. Consider the Banach space $L^p_k := \{f \in L^p(\mathbb{R}^d) : f \text{ is } G_k\text{-symmetric}\}$, whose corresponding optimal Stein–Tomas constant is

$$\mathbf{T}_{d,p}(k) := \sup_{0\neq f\in L^p_k} \frac{\left(\int_{\mathbb{S}^{d-1}} |\widehat{f}(\omega)|^2 \mathrm{d}\sigma(\omega)\right)^{\frac{1}{2}}}{\|f\|_{L^p(\mathbb{R}^d)}}. \tag{5}$$

Since $\mathbf{T}_{d,p}(d-k) = \mathbf{T}_{d,p}(k)$, we let $m := (d-k) \wedge k$ and note that:

- if $m = 0$, then we are dealing with radial functions, for which the restriction conjecture holds as seen in Sect. 1.1;
- if $m = 1$, then Knapp's example from Sect. 1.2 precludes any improvement over the exponent range (2).

What about $m \ge 2$? This will be addressed next.

2.1 Main Results

The results in this section were obtained in [8].

Theorem 1 ([8]) *Let $d \ge 4$ and $2 \le k \le d-2$ and $m = (d-k) \wedge k$. Then the inequality*

$$\int_{\mathbb{S}^{d-1}} |\widehat{f}(\omega)|^2 \,\mathrm{d}\sigma(\omega) \lesssim_{k,p,d} \|f\|^2_{L^p(\mathbb{R}^d)} \tag{6}$$

holds for every G_k-symmetric $f : \mathbb{R}^d \to \mathbb{C}$ if $1 \le p \le 2\frac{d+m}{d+m+2}$.

Estimate (6) improves the classical Stein–Tomas inequality since $2\frac{d+m}{d+m+2} > 2\frac{d+1}{d+3}$ if $m > 1$. The exponent range is optimal, as can be verified with a G_k-symmetric Knapp example; see Sect. 2.2 below. The proof of Theorem 1 is long and technical, relying on weighted two-dimensional restriction-type estimates for the operator

$$\int_{\mathbb{R}^2} \mathbf{1}_{|\omega_1||x_1|\ge 1} \mathbf{1}_{|\omega_2||x_2|\ge 1} (1+|x_1|)^{-\alpha}(1+|x_2|)^{-\beta} e^{-ix\cdot(|\omega_1|,|\omega_2|)} f(x)\,\mathrm{d}x.$$

In turn, these are accomplished via a careful use of the inequalities of Pitt and Stein–Weiss, which are weighted versions of the inequalities of Hausdorff–Young and Hardy–Littlewood–Sobolev.

Theorem 2 ([8]) *Let $d \geq 4$ and $2 \leq k \leq d-2$ and $m = (d-k) \wedge k \neq \frac{d}{2}$. Then the weak-type inequality*

$$\|\widehat{f}\|_{L^{\frac{2m}{m-1},\infty}(\mathbb{S}^{d-1})} \lesssim_{k,d} \|f\|_{L^{\frac{2m}{m+1},1}(\mathbb{R}^d)} \tag{7}$$

holds for every G_k-symmetric function $f : \mathbb{R}^d \to \mathbb{C}$.

In the remaining case when d is even and $m = \frac{d}{2}$, then estimate (7) has the following counterpart:

$$\|\widehat{f}\|_{L^{\frac{2d}{d-2},\infty}(\mathbb{S}^{d-1})} \lesssim_d \|f\|_{X_{\frac{2d}{d+2}}},$$

where $X_p := L^{p,1}_x(\mathbb{R}^m; L^{p,1}_y(\mathbb{R}^m)) + L^{p,1}_y(\mathbb{R}^m; L^{p,1}_x(\mathbb{R}^m))$ is the sum of two mixed Lorentz spaces. Via real interpolation methods for mixed Lorentz spaces that were very recently developed [6], we obtain the Riesz diagram in Fig. 2, which should be compared to the one in Fig. 1. We remark that estimates in the red region remain an enticing open question.

2.2 G_k-Symmetric Knapp

That no estimates hold in the gray region of Fig. 2 can be verified with a G_k-symmetric version of Knapp's example, to which we now turn our attention. First

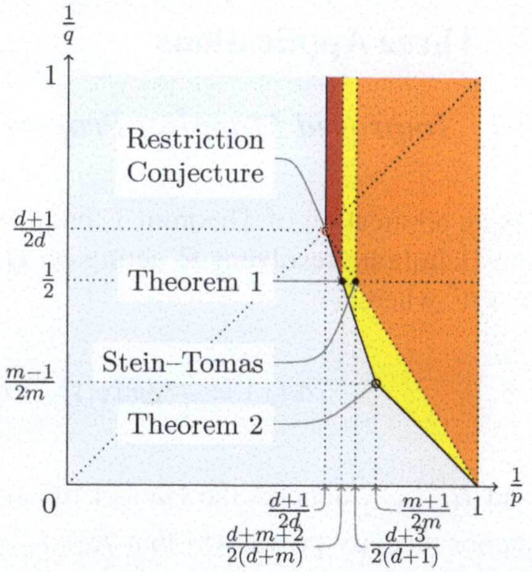

Fig. 2 Riesz diagram for the G_k-symmetric restriction operator to $\mathbb{S}^{d-1}$

of all, specializing (4) to the constant function $f \equiv \mathbf{1}$ yields

$$\widehat{\sigma}(x) \simeq_d |x|^{-\frac{d-1}{2}} J_{\frac{d-1}{2}}(|x|). \tag{8}$$

Let g denote the indicator function of the G_k-symmetric cap given by

$$\mathcal{C}_{\delta,k} = \{(\eta, \zeta) \in \mathbb{R}^{d-k} \times \mathbb{R}^k : |\eta|^2 + |\zeta|^2 = 1,\ |\eta| < \delta\}.$$

A straightforward computation in spherical coordinates then yields $\|g\|_{L^{q'}(\mathbb{S}^{d-1})} \simeq \delta^{\frac{d-k}{q'}}$. On the other hand, writing $x = (y, z) \in \mathbb{R}^{d-k} \times \mathbb{R}^k$, appealing to (8) on $\mathbb{S}^{d-k-1}$ and $\mathbb{S}^{k-1}$, and carefully pointwise estimating the extension operator

$$\widehat{g\sigma}(x) = \int_0^\delta r^{d-k-1}(1-r^2)^{\frac{k-2}{2}} \frac{J_{\frac{d-k-2}{2}}(r|y|)}{(r|y|)^{\frac{d-k-2}{2}}} \frac{J_{\frac{k-2}{2}}(\sqrt{1-r^2}|z|)}{(\sqrt{1-r^2}|z|)^{\frac{k-2}{2}}}\, \mathrm{d}r$$

yields $|\widehat{g\sigma}(x)| \gtrsim \delta^{d-k} j^{\frac{1-k}{2}}$ if $x \in E_j$, where $E_j := \{x \in \mathbb{R}^d : 0 \le |y| \le c\delta^{-1}, \frac{z_j - c}{\sqrt{1-\delta^2}} \le |z| \le z_j + c\}$ and $\{z_j\}_{j\ge 1}$ is the sequence of local maxima of the Bessel function $J_{\frac{k-2}{2}}$. All in all, we then have that

$$\frac{\|\widehat{g\sigma}\|_{L^{p'}(\mathbb{R}^d)}}{\|g\|_{L^{q'}(\mathbb{S}^{d-1})}} \gtrsim \begin{cases} \delta^{\frac{d+k}{p} + \frac{d-k}{q} - d - 1}, & \text{if } \frac{1}{p} < \frac{k+1}{2k}, \\ \delta^{\frac{d+k}{p} + \frac{d-k}{q} - d - 1} |\log(\delta)|^{\frac{1}{p'}}, & \text{if } \frac{1}{p} = \frac{k+1}{2k}, \\ \delta^{\frac{d-k}{p} + \frac{d-k}{q} - d + k}, & \text{if } \frac{1}{p} > \frac{k+1}{2k}, \end{cases}$$

from where the claim follows.

3 Three Applications

3.1 Improved Mapping Properties of Helmholtz Resolvent

As an application of Theorem 1, one may deduce improved mapping properties of the Helmholtz resolvent $\mathcal{R}$ acting on G_k-symmetric functions. Recall that $\mathcal{R}f = f * \Phi$, where

$$\Phi(x) = \frac{i}{4}(2\pi|x|)^{\frac{2-d}{2}} H^{(1)}_{\frac{d-2}{2}}(|x|) \text{ for } x \in \mathbb{R}^d \setminus \{0\},$$

and $H^{(1)}_{(d-2)/2}$ denotes the Hankel function of the first kind of order $\frac{d-2}{2}$. The first author recently proved [5] that $\mathcal{R} : L^p_k \to L^q_k$ is bounded provided that $1 \le k \le$

$d-1$ and

$$\frac{1}{p}\wedge\frac{1}{q'} > \frac{d-1}{2d}, \qquad \frac{1}{p}-\frac{1}{q} \geq \frac{2}{d+m}, \qquad (m := (d-k)\wedge k).$$

In the special case $q = p'$, this leads to the G_k-symmetric Stein–Tomas exponent $p = 2\frac{d+m}{d+m+2}$, which is in fact a crucial ingredient of the proof. This should be compared to the work of Kenig–Ruiz–Sogge [2] in the non-symmetric setting.

3.2 *Unconditional Existence of Endpoint Stein–Tomas Maximizers*

The question of whether maximizers for the endpoint Stein–Tomas inequality exist has recently received a great deal of attention [10]. In particular, Frank–Lieb–Sabin [1] answered it in the affirmative in all dimensions, but only under the assumption that Lieb's conjecture[3] holds.

In the G_k-symmetric setting, we have the following unconditional result. Recall that the constrained optimal constant $\mathbf{T}_{d,p}(k)$ was defined in (5).

Theorem 3 ([8]) *Let $d \geq 4$ and $2 \leq k \leq d-2$ and $1 \leq p < 2\frac{d+m}{d+m+2}$. Maximizing sequences for $\mathbf{T}_{d,p}(k)$, normalized in $L^p(\mathbb{R}^d)$, are precompact in L^p_k.*

In particular, maximizers for $\mathbf{T}_{d,k}(p)$ exist at the Stein–Tomas endpoint $p = 2\frac{d+1}{d+3}$. The key lies in the following result which, in the language of concentration-compactness, precludes *vanishing* of maximizing sequences.

Lemma 4 ([8]) *Let $d \geq 4$ and $2 \leq k \leq d-2$. Then, for every $\varepsilon > 0$, there exists $\rho = \rho(k, d, \varepsilon) < \infty$ for which $|\widehat{f\sigma}(x)| < \varepsilon$ if $|x| > \rho$, for every G_k-symmetric $f \in L^2(\mathbb{S}^{d-1})$ such that $\|f\|_{L^2(\mathbb{S}^{d-1})} = 1$.*

Note that no such property can hold for general $f \in L^2(\mathbb{S}^{d-1})$ since the extension operator interchanges modulations and translations: given $y \in \mathbb{R}^d$,

$$[(e^{iy\cdot} f)\sigma]^\wedge(x) = \int_{\mathbb{S}^{d-1}} e^{iy\cdot\omega} f(\omega) e^{ix\cdot\omega}\, \mathrm{d}\sigma(\omega) = \widehat{f\sigma}(x+y).$$

On the other hand, if a nonzero function f and its modulation $e^{iy\cdot} f$ are both G_k-symmetric on $\mathbb{S}^{d-1}$, then necessarily $y = 0$. Set $2_\star := 2\frac{d+1}{d-1}$ and $2_\star^k := 2\frac{d+m}{d-2+m}$ where $m = (d-k)\wedge k$. Since $2_\star^k < 2_\star < \infty$ if $2 \leq k \leq d-2$, there exists $\theta \in (0, 1)$

[3] This is the statement that gaussians maximize the Stein–Tomas extension inequality on the paraboloid, or equivalently, the Strichartz estimate for the Schrödinger equation; see [1] for further discussion.

such that any maximizing sequence $(f_n) \subset L^p_k$ for $\mathbf{T}_{d,p}(k)$ satisfies

$$\tfrac{1}{2}\mathbf{T}_{d,p}(k) \le \|\widehat{f_n\sigma}\|_{2_\star} \le \|\widehat{f_n\sigma}\|^{\theta}_{2^k_\star}\|\widehat{f_n\sigma}\|^{1-\theta}_{\infty}$$

if n is large enough, and so there exists a sequence $(x_n) \subset \mathbb{R}^d$ such that $|\widehat{f_n\sigma}(x_n)| \ge \varepsilon_0 > 0$. Lemma 4 then implies that the sequence (x_n) is bounded, and this leads (via the compactness theorem of Arzelà–Ascoli) to the existence of a nonzero weak limit. The result follows.

3.3 Low Energy Solutions of Biharmonic NLS

The results in this section were obtained in [7]. Another fashionable topic is that of biharmonic nonlinear Schrödinger equations (NLS). A concrete model is

$$\Delta^2 u + 2a\Delta u + bu = |u|^{p-2}u \quad \text{in } \mathbb{R}^d, \tag{9}$$

where $a, b > 0$ are positive constants. Defining the quotient

$$R_{a,b}(p) := \inf_{0\neq u\in H^2(\mathbb{R}^d)} \frac{q_{a,b}(u)}{\|u\|_p^2}, \tag{10}$$

where the quadratic form $q_{a,b}$ is given by

$$q_{a,b}(u) = \int_{\mathbb{R}^d} (|\Delta u|^2 - 2a|\nabla u|^2 + b|u|^2)\, \mathrm{d}x = \int_{\mathbb{R}^d} g_{a,b}(|\xi|)|\widehat{u}(\xi)|^2\, \mathrm{d}\xi, \tag{11}$$

we are interested in *ground states*, i.e., nonzero solutions $u \in H^2(\mathbb{R}^d)$ of (9) for which the infimum in (10) is attained. The second identity in (11) follows from Plancherel's theorem provided one sets $g_{a,b}(r) := r^4 - 2ar^2 + b = (r^2 - a)^2 + b - a^2$. Two observations are in order:

- if $a \le 0$, then $g_{a,b}(r)$ strictly increases with $r \ge 0$, and so every ground state of (9) is radial if p is even. This follows from the Fourier rearrangement techniques recently developed by Lenzmann–Sok [3];
- if $a > 0$, then $g_{a,b}(r)$ is not monotonic on $\mathbb{R}_+$, and the methods of [3] do not apply.

We focus on the latter case, $a > 0$. By scaling, it suffices to consider $(a, b) = (1, 1 + \varepsilon)$ for some $\varepsilon > 0$, the corresponding equation being

$$\Delta^2 u + 2\Delta u + (1 + \varepsilon)u = |u|^{p-2}u \quad \text{in } \mathbb{R}^d \tag{12}$$

with quotient $\mathrm{R}^{\circ}_{\varepsilon}(p) := \inf_{0\neq u\in H^2(\mathbb{R}^d)} \|u\|_p^{-2}\mathrm{q}_\varepsilon(u)$, where

$$\mathrm{q}_\varepsilon(u) = \int_{\mathbb{R}^d} \left(|\Delta u|^2 - 2|\nabla u|^2 + (1+\varepsilon)|u|^2\right) \mathrm{d}x = \int_{\mathbb{R}^d} g_\varepsilon(|\xi|)|\widehat{u}(\xi)|^2 \,\mathrm{d}\xi$$

and $g_\varepsilon(r) = (r^2-1)^2 + \varepsilon$. Lenzmann–Weth [4] established the radial symmetry breaking of ground states for (12). The following is their precise result.[4]

Theorem 5 ([4]) *Let $d \geq 2$ and $2 < p < 2^1_\star$. There exists $\varepsilon_0 = \varepsilon_0(p,d) > 0$ such that every ground state of* (12) *is nonradial if* $0 < \varepsilon < \varepsilon_0$.

To analyze the more delicate possibility of G_k-symmetry breaking, we define the quantities

$$\mathrm{R}^k_\varepsilon(p) := \inf_{0\neq u\in H^2_k} \frac{\mathrm{q}_\varepsilon(u)}{\|u\|_p^2}, \qquad \mathrm{R}^{\mathrm{rad}}_\varepsilon(p) := \inf_{0\neq u\in H^2_{\mathrm{rad}}} \frac{\mathrm{q}_\varepsilon(u)}{\|u\|_p^2}$$

where $H^2_k := \{f \in H^2(\mathbb{R}^d) : f \text{ is } G_k\text{-symmetric}\}$ and $H^2_{\mathrm{rad}} := \{f \in H^2(\mathbb{R}^d) : f \text{ is radial}\}$ and have the following result.

Theorem 6 ([7]) *Let $d \geq 2$ and $1 \leq k \leq \lfloor d/2 \rfloor$ and $2 < p < 2^k_\star$. There exists $\varepsilon_0 = \varepsilon_0(p,d,k)$ such that, for $0 < \varepsilon < \varepsilon_0$, the following chain of strict inequalities holds:*

$$\mathrm{R}^{\circ}_\varepsilon(p) \vee \mathrm{R}^1_\varepsilon(p) < \mathrm{R}^2_\varepsilon(p) < \cdots < \mathrm{R}^k_\varepsilon(p) < \mathrm{R}^{k+1}_\varepsilon(p) \wedge \ldots \wedge \mathrm{R}^{\lfloor d/2\rfloor}_\varepsilon(p)$$
$$\leq \mathrm{R}^{\mathrm{rad}}_\varepsilon(p) < \infty.$$

Moreover, each infimum is attained at some nontrivial solution of (12), *except possibly* $\mathrm{R}^1_\varepsilon(p)$.

In particular, there exist $k+1$ different nontrivial solutions and the ground state, i.e., the minimizer of the quotient $\mathrm{R}^{\circ}_\varepsilon(p)$, is neither radial nor G_k-symmetric. Note that the case $k = 1$ of Theorem 6 already implies Theorem 5.

Acknowledgments DOS is partially supported by IST Santander Start Up Funds and Centro de Análise Matemática, Geometria e Sistemas Dinâmicos (CAMGSD). He expresses his gratitude to Roland Duduchava, Eugene Shargorodsky, George Tephnadze, Medea Tsaava, Giorgi Tutberidze and Zurab Vashakidze for organizing the Tbilisi Analysis & PDE Workshop 2023 at The University of Georgia, whose hospitality is greatly appreciated.

[4] Recall that $2^1_\star = 2_\star = 2\frac{d+1}{d-1}$ is the endpoint exponent for the Stein–Tomas inequality in extension form.

References

1. R. Frank, E.H. Lieb, J. Sabin, Maximizers for the Stein–Tomas inequality. Geom. Funct. Anal. **26**(4), 1095–1134 (2016)
2. C. Kenig, A. Ruiz, C. Sogge, Uniform Sobolev inequalities and unique continuation for second order constant coefficient differential operators. Duke Math. J. **55**(2), 329–347 (1987)
3. E. Lenzmann, J. Sok, A sharp rearrangement principle in Fourier space and symmetry results for PDEs with arbitrary order. Int. Math. Res. Not. IMRN **2021**(19), 15040–15081 (2021)
4. E. Lenzmann, T. Weth, Symmetry breaking for ground states of biharmonic NLS via Fourier extension estimates. JAMA **152**, 777–800 (2024)
5. R. Mandel, The restriction-extension operator on Lebesgue spaces with symmetries and applications to PDEs. arXiv:2303.03020
6. R. Mandel, Real interpolation for mixed Lorentz spaces and Minkowski's inequality. Z. Anal. Anwend. **42**(3/4), 457–469 (2023)
7. R. Mandel, D. Oliveira e Silva, Block-radial symmetry breaking for ground states of biharmonic NLS. Calc. Var. Partial Differential Equations **63**(2), 20 (2024)
8. R. Mandel, D. Oliveira e Silva, The Stein–Tomas inequality under the effect of symmetries. J. Anal. Math. **150**(2), 547–582 (2023)
9. C. Muscalu, W. Schlag, *Classical and Multilinear Harmonic Analysis*. Vol. I. Cambridge Stud. Adv. Math., vol. 137 (Cambridge University Press, Cambridge, 2013)
10. G. Negro, D. Oliveirae Silva, C. Thiele, When doese $e^{-|\tau|}$ maximize Fourier extension for a conic section? Adv. Anal. Geom., vol. 9, 391–426 (De Gruyter, Berlin, 2023)
11. E.M. Stein, *Harmonic Analysis: Real-Variable Methods, Orthogonality, and Oscillatory Integrals* (Princeton University Press, Princeton, 1993)
12. P. Tomas, A restriction theorem for the Fourier transform. Bull. Amer. Math. Soc. **81**, 477–478 (1975)

On Solonnikov Parabolicity of the Evolution Anisotropic Stokes and Oseen PDE Systems

Sergey E. Mikhailov

Abstract In this note it is shown that the evolution (non-stationary) anisotropic Stokes and generalised Oseen systems with variable viscosity coefficients in a compressible framework can be classified as parabolic PDE systems in the sense of Solonnikov.

Keywords Partial differential equations · Evolution anisotropic Stokes and Oseen equations · Parabolicity

2000 Mathematics Subject Classification 35K40, 35Q30, 76D07

1 Introduction: Solonnikov Parabolicity

In this section, we present a generalised notion of parabolicity for PDE systems introduced by Solonnikov in [7, Section 1] (see also [5, Chapter VII, Section 8, Definition 4], [2, Definition I.4]). It can be considered as a parabolic counterpart of the Agmon-Douglis-Nirenberg ellipticity for PDE systems.

Following [5, Chapter VII, Section 8], we will begin with the Petrovskii definition of a scalar parabolic equation. Let $L(x, t, \partial/\partial\mathbf{x}, \partial/\partial t)$ be a scalar linear partial differential operator of arbitrary order with complex coefficients depending on $\mathbf{x}$ and t. Let $\boldsymbol{\xi} \in \mathbb{R}^n$ be an n-dimensional vector and $\varrho \in \mathbb{C}$ be a scalar. It is clear that at any point $(\mathbf{x}, t)$, the function $L(\mathbf{x}, t, i\boldsymbol{\xi}, \varrho)$ is a polynomial in ξ_j and ϱ. Let b be some positive integer and let the degree of the polynomial $L(\mathbf{x}, t, i\boldsymbol{\xi}\lambda, \varrho\lambda^{2b})$ in λ be equal to $2br$, where r is a positive integer. We denote by L_0 the principal part of the polynomial L, i.e., the sum of those terms of L, for which

$$L_0(\mathbf{x}, t, i\boldsymbol{\xi}\lambda, \varrho\lambda^{2b}) = \lambda^{2br} L_0(\mathbf{x}, t, i\boldsymbol{\xi}, \varrho). \tag{1.1}$$

S. E. Mikhailov (✉)
Brunel University London, London, UK
e-mail: sergey.mikhailov@brunel.ac.uk

R. Duduchava et al. (eds.), *Tbilisi Analysis and PDE Seminar*,
Research Perspectives Ghent Analysis and PDE Center 7,
https://doi.org/10.1007/978-3-031-62894-8_15

Definition 1.1 *A scalar partial differential operator L is said to be parabolic (2b-parabolic) at a point $(\mathbf{x}, t)$ if for any real vector $\boldsymbol{\xi}$ the roots, ϱ_s, of the principal part, $L_0(\mathbf{x}, t, i\boldsymbol{\xi}, \varrho)$, of the polynomial L in variable ϱ satisfy the condition*

$$\operatorname{Re} \varrho_s \le -\delta |\boldsymbol{\xi}|^{2b} \tag{1.2}$$

for some constant $\delta > 0$.

Definition 1.2 *Let $m \ge 2$ be a positive integer. A matrix differential operator $\mathbb{L}(\mathbf{x}, t, \partial_{\mathbf{x}}, \partial_t)$ with elements $\mathbb{L}_{kj}(\mathbf{x}, t, \partial_{\mathbf{x}}, \partial_t)$, $(k, j = 1, \ldots, m)$ will be called parabolic at point $(\mathbf{x}, t)$ in the sense of Solonnikov if:*

(i) the operator

$$L(x, t, \partial/\partial\mathbf{x}, \partial/\partial t) = \det \mathbb{L}(\mathbf{x}, t, \partial_{\mathbf{x}}, \partial_t)$$

is 2b-parabolic in the sense of Definition 1.1 for some positive integer b;

(ii) there exist integers s_ℓ and t_ℓ, $\boldsymbol{\ell} = 1, \ldots, m$ such that the degree of the polynomial $\mathbb{L}_{\boldsymbol{\ell} j}(\mathbf{x}, t, i\boldsymbol{\xi}\lambda, \varrho\lambda^{2b})$ in λ does not exceed $s_\ell + t_j$ (if $s_\ell + t_j < 0$ then $\mathbb{L}_{\boldsymbol{\ell} j} = 0$) and, in addition,

$$\sum_{\ell=1}^{m} (s_\ell + t_\ell) = 2br,$$

where r is the degree of the polynomial $L(\mathbf{x}, t, i\boldsymbol{\xi}, \varrho)$ in the variable ϱ.

2 Anisotropic Stokes and Oseen PDE Systems

Let $n \ge 2$ be an integer, $\mathbf{x} \in \mathbb{R}^n$ denote the space coordinate vector, and $t \in \mathbb{R}$ be time variable. Let $\mathfrak{L}$ denote the second-order differential operator represented in the component-wise divergence form as

$$(\mathfrak{L}\mathbf{u})_k := \partial_\alpha \big(a_{kj}^{\alpha\beta} E_{j\beta}(\mathbf{u}) \big), \quad k = 1, \ldots, n, \tag{2.1}$$

where $\mathbf{u} = (u_1, \ldots, u_n)^\top$, $E_{j\beta}(\mathbf{u}) := \frac{1}{2}(\partial_j u_\beta + \partial_\beta u_j)$ are the entries of the symmetric part $\mathbb{E}(\mathbf{u})$ of the gradient, $\nabla\mathbf{u}$, of $\mathbf{u}$ in space coordinates, and $a_{kj}^{\alpha\beta}(\mathbf{x}, t)$ are variable components of the tensor viscosity coefficient, cf. [1], $\mathbb{A}(\mathbf{x}, t) = \left\{ a_{kj}^{\alpha\beta}(\mathbf{x}, t) \right\}_{i,j,\alpha,\beta=1}^{n}$ depending on the space coordinate vector $\mathbf{x}$ and time t. We also denoted $\partial_j = \dfrac{\partial}{\partial x_j}$. Here and further on, the Einstein convention on summation in repeated indices from 1 to n is used unless stated otherwise.

The following symmetry conditions are assumed (cf. [6, Eq. (3.1), (3.3)]),

$$a_{kj}^{\alpha\beta}(\mathbf{x},t) = a_{\alpha j}^{k\beta}(\mathbf{x},t) = a_{k\beta}^{\alpha j}(\mathbf{x},t). \tag{2.2}$$

In addition, we require that the tensor $\mathbb{A}$ satisfies *the relaxed ellipticity condition* in terms of all *symmetric* matrices in $\mathbb{R}^{n\times n}$ with *zero matrix trace*, see [3, 4]. Thus, we assume that there exists a constant $C_{\mathbb{A}} > 0$ such that,

$$a_{kj}^{\alpha\beta}(\mathbf{x},t)\zeta_{k\alpha}\zeta_{j\beta} \geq C_{\mathbb{A}}^{-1}|\boldsymbol{\zeta}|^2, \quad \text{for a.e. } \mathbf{x}, t, \tag{2.3}$$

$$\forall\, \boldsymbol{\zeta} = \{\zeta_{k\alpha}\}_{k,\alpha=1,\dots,n} \in \mathbb{R}^{n\times n} \text{ such that } \boldsymbol{\zeta} = \boldsymbol{\zeta}^{\top} \text{ and } \sum_{k=1}^{n}\zeta_{kk} = 0,$$

where $|\boldsymbol{\zeta}|^2 = \zeta_{k\alpha}\zeta_{k\alpha}$, and the superscript $^\top$ denotes the transpose of a matrix. Note that in the more common, strong ellipticity condition, inequality (2.3) should be satisfied for all matrices (not only symmetric with zero trace), which makes it much more restrictive.

We assume that $a_{ij}^{\alpha\beta} \in L_\infty(\mathbb{R}^n \times [0,T])$, where $[0,T]$ is some time interval, and the tensor $\mathbb{A}$ is endowed with the norm

$$\|\mathbb{A}\| := \max\left[\|a_{ij}^{\alpha\beta}\|_{L_\infty(\mathbb{R}^n\times[0,T])} : k, j, \alpha, \beta \in \{1\dots,n\}\right] < \infty. \tag{2.4}$$

Symmetry conditions (2.2) lead to the following equivalent form of the operator $\mathfrak{L}$

$$(\mathfrak{L}\mathbf{u})_k = \partial_\alpha\big(a_{kj}^{\alpha\beta}\partial_\beta u_j\big), \quad k = 1,\dots,n. \tag{2.5}$$

Let $\mathbf{u}(\mathbf{x},t)$ be an unknown vector velocity field, $p(\mathbf{x},t)$ be an unknown (scalar) pressure field, $\mathbf{f}(\mathbf{x},t)$ be a given vector field and $g(\mathbf{x},t)$ be a given scalar field defined in $\mathbb{R}^n$, where $t \in \mathbb{R}$ is the time variable. Then the linear PDE system

$$\partial_t\mathbf{u}(\mathbf{x},t) - \mathfrak{L}\mathbf{u}(\mathbf{x},t) + \nabla p(\mathbf{x},t) = \mathbf{f}(\mathbf{x},t), \tag{2.6}$$

$$\operatorname{div}\mathbf{u}(\mathbf{x},t) = g(\mathbf{x},t). \tag{2.7}$$

determines the *anisotropic evolution Stokes system in a compressible framework*.

A more general linear system

$$\partial_t\mathbf{u}(\mathbf{x},t) - \mathfrak{L}\mathbf{u}(\mathbf{x},t) + \nabla p(\mathbf{x},t) + (\mathbf{U}(\mathbf{x},t)\cdot\nabla)\mathbf{u}(\mathbf{x},t) = \mathbf{f}(\mathbf{x},t), \tag{2.8}$$

$$\operatorname{div}\mathbf{u}(\mathbf{x},t) = g(\mathbf{x},t), \tag{2.9}$$

where $\mathbf{U}(\mathbf{x},t)$ is a given vector field defined at a.e. $(\mathbf{x},t)$, is the (generalised) evolution Oseen system.

If $g = 0$ in (2.7) and (2.9), then systems (2.6)–(2.7) and (2.8)–(2.9) are reduced to the evolution *incompressible* anisotropic Stokes and Oseen systems, respectively.

In the *isotropic case*, the tensor $\mathbb{A}$ reduces to

$$a_{kj}^{\alpha\beta}(\mathbf{x},t) = \lambda(\mathbf{x},t)\delta_{k\alpha}\delta_{j\beta} + \mu(\mathbf{x},t)\left(\delta_{\alpha j}\delta_{\beta k} + \delta_{\alpha\beta}\delta_{kj}\right), \quad 1 \le k, j, \alpha, \beta \le n, \tag{2.10}$$

where $\lambda, \mu \in L_\infty(\mathbb{R}^n \times [0,T])$, and $c_\mu^{-1} \le \mu(\mathbf{x},t) \le c_\mu$ for a.e. $\mathbf{x}$ and t, with some constant $c_\mu > 0$ (cf., e.g., Appendix III, Part I, Section 1 in [8]). Then it is immediate that condition (2.3) is fulfilled with $C_{\mathbb{A}} = c_\mu/2$ and thus our results apply also to the Stokes and Oseen systems in the *isotropic case*. Moreover, (2.1) becomes

$$\mathfrak{L}\mathbf{u} = (\lambda+\mu)\nabla\operatorname{div}\mathbf{u} + \mu\Delta\mathbf{u} + (\nabla\lambda)\operatorname{div}\mathbf{u} + 2(\nabla\mu)\cdot\mathbb{E}(\mathbf{u}) \tag{2.11}$$

Assuming $\lambda = 0$ and $\mu = 1$ we arrive at the classical mathematical formulations of isotropic, constant-coefficient Stokes and Oseen systems.

3 Classification of the Evolution Anisotropic Stokes and Oseen Systems

Let us consider the Oseen system (2.8)–(2.9). It can be re-written as

$$\mathbf{A}_{\mathrm{U}}\begin{pmatrix}\mathbf{u}\\ p\end{pmatrix} = \begin{pmatrix}\mathbf{f}\\ g\end{pmatrix}, \tag{3.1}$$

where the operator $\mathbf{A}_{\mathrm{U}} = \mathbf{A}_{\mathrm{U}}(\mathbf{x}, t, \partial_{\mathbf{x}}, \partial_t)$ is defined as

$$\mathbf{A}_{\mathrm{U}}\begin{pmatrix}\mathbf{u}\\ p\end{pmatrix} := \begin{pmatrix}\partial_t\mathbf{u} - \mathfrak{L}\mathbf{u} + (\mathbf{U}\cdot\nabla)\mathbf{u} + \nabla p\\ \operatorname{div}\mathbf{u}\end{pmatrix}. \tag{3.2}$$

If $\mathbf{U} \equiv \mathbf{0}$, the evolution Oseen system (2.8)–(2.9) reduces to the evolution Stokes system (2.6)–(2.7) that can be re-written as

$$\mathbf{A}\begin{pmatrix}\mathbf{u}\\ p\end{pmatrix} = \begin{pmatrix}\mathbf{f}\\ g\end{pmatrix}, \tag{3.3}$$

where the operator $\mathbf{A} = \mathbf{A}(\mathbf{x}, t, \partial_{\mathbf{x}}, \partial_t)$ is defined as

$$\mathbf{A}\begin{pmatrix}\mathbf{u}\\ p\end{pmatrix} = \begin{pmatrix}\partial_t\mathbf{u} - \mathfrak{L}\mathbf{u} + \nabla p\\ \operatorname{div}\mathbf{u}\end{pmatrix}. \tag{3.4}$$

Let us classify the operator $\mathbf{A}$ and the PDE system (2.6)–(2.7). Since Eq. (2.7) does not include the time derivative, ∂_t, PDE system (2.6)–(2.7) and the operator

$\mathbf{\Lambda}$ are not parabolic in the sense of Petrovskii or in the sense of Shirota (see the corresponding definition, e.g., in [5, Chapter VII, Section 8, Definitions 2 and 3], [2, Definitions I.2 and I.5]). However we will prove that under the condition (2.3) of relaxed ellipticity for the viscosity tensor, the operator $\mathbf{\Lambda}$ is parabolic in the sense of Solonnikov.

To this end, we first need to prove several auxiliary results. Replacing in the operator $\mathbf{\Lambda}$ arguments ∂_j by $i\xi_j$ and ∂_t by ρ we express its matrix entries as polynomials in $\rho \in \mathbb{C}$ and $\boldsymbol{\xi} \in \mathbb{R}^n$, i.e.,

$$\Lambda_{\ell j}(\mathbf{x}, t, i\boldsymbol{\xi}, \rho) = \begin{cases} \rho\delta_{\ell j} + \xi_\alpha a_{\ell j}^{\alpha\beta}\xi_\beta, & \ell, j = 1, \ldots, n; \\ -i\xi_\ell, & \ell = 1, \ldots, n, \ j = n+1; \\ -i\xi_j, & \ell = n+1, \ j = 1, \ldots, n; \\ 0, & \ell = j = n+1. \end{cases} \tag{3.5}$$

Here $a_{\ell j}^{\alpha\beta} = a_{\ell j}^{\alpha\beta}(\mathbf{x}, t)$.

Lemma 3.1 *Let $n \geq 2$ and the relaxed ellipticity condition* (2.3) *hold at a point* $(\mathbf{x}, t)$. *Let $\rho \in \mathbb{C}$ and $\boldsymbol{\xi} \in \mathbb{R}^n$. If* $\det \mathbf{\Lambda}(\mathbf{x}, t, i\boldsymbol{\xi}, \rho) = 0$, *then*

$$\operatorname{Re} \rho \leq -\frac{1}{2} C_{\mathbb{A}}^{-1} |\boldsymbol{\xi}|^2. \tag{3.6}$$

Proof Let us assume that $\det \mathbf{\Lambda}(\mathbf{x}, t, i\boldsymbol{\xi}, \rho) = 0$. If $\boldsymbol{\xi} = \mathbf{0}$, then from (3.5) we have $\rho = 0$ implying that inequality (3.6) is satisfied.

Let us now consider the case $\boldsymbol{\xi} \neq \mathbf{0}$. Then there should exist a non-trivial vector $(\mathbf{w}, q) \in \mathbb{C}^n \times \mathbb{C}$ solving the homogeneous algebraic system

$$\mathbf{\Lambda}(\mathbf{x}, t, i\boldsymbol{\xi}, \rho)\begin{pmatrix}\mathbf{w}\\ q\end{pmatrix} = \begin{pmatrix}\mathbf{0}\\ 0\end{pmatrix}. \tag{3.7}$$

If we assume $\mathbf{w} = \mathbf{0}$, then the first n equations of (3.7) imply that $\boldsymbol{\xi} q = \mathbf{0}$, and since $\boldsymbol{\xi} \neq \mathbf{0}$, this means $q = 0$, that is, the vector $(\mathbf{w}, q)$ is trivial. Then we further assume that $\mathbf{w} \neq \mathbf{0}$. The last equation of the system implies

$$\mathbf{w} \cdot \boldsymbol{\xi} = w_j \xi_j = 0. \tag{3.8}$$

After the scalar multiplication of the system (3.7) by the vector $(\overline{\mathbf{w}}, 0)$, where the bar means complex conjugate, we have,

$$\overline{w}_\ell[\Lambda_{\ell j}(\mathbf{x}, t, i\xi, \rho)w_j - i\xi_\ell q] = \rho|\mathbf{w}|^2 + a_0(\mathbf{w}, \mathbf{w}) = 0, \tag{3.9}$$

where

$$a_0(\mathbf{w},\mathbf{w}) := \overline{w}_\ell \xi_\alpha a_{\ell j}^{\alpha\beta} \xi_\beta w_j = a_{\ell j}^{\alpha\beta} (\overline{\mathbf{w}} \otimes \boldsymbol{\xi})_{\ell\alpha} (\mathbf{w} \otimes \boldsymbol{\xi})_{\beta j}$$
$$= a_{\ell j}^{\alpha\beta} (\overline{\mathbf{w}} \otimes \boldsymbol{\xi})^s_{\ell\alpha} (\mathbf{w} \otimes \boldsymbol{\xi})^s_{\beta j}. \tag{3.10}$$

due to the symmetry conditions (2.2). Here $(\mathbf{w} \otimes \boldsymbol{\xi})^s$ is the symmetric part of the matrix $\mathbf{w} \otimes \boldsymbol{\xi}$, i.e.,

$$(\mathbf{w} \otimes \boldsymbol{\xi})^s_{\ell\alpha} := \frac{1}{2}(w_\ell \xi_\alpha + w_\alpha \xi_\ell), \quad \ell, \alpha = 1, \ldots, n. \tag{3.11}$$

By (3.8) and the ellipticity condition (2.3), Eq. (3.9) implies

$$\begin{aligned}
-\operatorname{Re}\rho|\mathbf{w}|^2 &= \operatorname{Re} a_0(\mathbf{w},\mathbf{w}) \\
&= a_{\ell j}^{\alpha\beta} (\operatorname{Re}\mathbf{w} \otimes \boldsymbol{\xi})^s_{\ell\alpha} (\operatorname{Re}\mathbf{w} \otimes \boldsymbol{\xi})^s_{\beta j} + a_{\ell j}^{\alpha\beta} (\operatorname{Im}\mathbf{w} \otimes \boldsymbol{\xi})^s_{\ell\alpha} (\operatorname{Im}\mathbf{w} \otimes \boldsymbol{\xi})^s_{\beta j} \\
&\geq C_{\mathbb{A}}^{-1} (|(\operatorname{Re}\mathbf{w} \otimes \boldsymbol{\xi})^s|^2 + |(\operatorname{Im}\mathbf{w} \otimes \boldsymbol{\xi})^s|^2) \\
&= \frac{1}{2} C_{\mathbb{A}}^{-1} (|\operatorname{Re}\mathbf{w}|^2 |\boldsymbol{\xi}|^2 + |\operatorname{Im}\mathbf{w}|^2 |\boldsymbol{\xi}|^2) = \frac{1}{2} C_{\mathbb{A}}^{-1} |\mathbf{w}|^2 |\boldsymbol{\xi}|^2,
\end{aligned}$$

which leads to (3.6). □

Let us introduce the differential operator $L(\mathbf{x}, t, \partial_\mathbf{x}, \partial_t) = \det \boldsymbol{\Lambda}(\mathbf{x}, t, \partial_\mathbf{x}, \partial_t)$. Then $L(\mathbf{x}, t, i\boldsymbol{\xi}, \rho) = \det \boldsymbol{\Lambda}(\mathbf{x}, t, i\boldsymbol{\xi}, \rho)$ is a polynomial in $\rho \in \mathbb{C}$ and $\boldsymbol{\xi} \in \mathbb{R}^n$ and moreover,

$$L(\mathbf{x}, t, i\boldsymbol{\xi}\lambda, \rho\lambda^2) = \lambda^{2n} L(\mathbf{x}, t, i\boldsymbol{\xi}, \rho). \tag{3.12}$$

Theorem 3.2 *Let $n \geq 2$ and the relaxed ellipticity condition* (2.3) *hold at a point* $(\mathbf{x}, t)$. *Then at this point the evolution anisotropic Stokes operator $\boldsymbol{\Lambda}$ defined by* (3.4) *is parabolic in the sense of Solonnikov, i.e., satisfies Definition 1.2.*

Proof We can choose $m = n + 1$ in Definition 1.2. By (3.12), the scalar operator $L(\mathbf{x}, t, \partial_\mathbf{x}, \partial_t) = \det \boldsymbol{\Lambda}(\mathbf{x}, t, \partial_\mathbf{x}, \partial_t)$ satisfies (1.1) with $L_0 = L$, $b = 1$, and $r = n$. Due to Lemma 3.1, then the operator $\det \boldsymbol{\Lambda}(\mathbf{x}, t, \partial_\mathbf{x}, \partial_t)$ satisfies Definition 1.1 with $\delta = \frac{1}{2} C_{\mathbb{A}}^{-1}$. Hence it is 2-parabolic and item (i) in Definition 1.2 is satisfied.

Due to (3.5), item (ii) in Definition 1.2 is satisfied for $s_\ell = t_\ell = 1$, $\ell = 1, \ldots, n$ and $s_{n+1} = t_{n+1} = 0$. Note that another choice, e.g., $s_\ell = 0$, $t_\ell = 2$, $\ell = 1, \ldots, n$ and $s_{n+1} = -1$, $t_{n+1} = 1$ is also possible. □

It is easy to see that the principal part of the Oseen operator (3.2) is obtained by taking $\mathbf{U} \equiv \mathbf{0}$, that is, it is given by the corresponding Stokes operator, which leads to the following assertion implied by Theorem 3.2.

Corollary 3.3 *Let* $n \geq 2$, *the relaxed ellipticity condition* (2.3) *hold at a point* $(\mathbf{x}, t)$ *and the function* $\mathbf{U}(\mathbf{x}, t)$ *be defined at* $(\mathbf{x}, t)$. *Then at this point the evolution anisotropic Oseen operator* $\mathbf{\Lambda}_{\mathrm{U}}$ *defined by* (3.2) *is parabolic in the sense of Solonnikov, i.e., satisfies Definition 1.2.*

The classification of the evolution anisotropic Stokes and Oseen operators as parabolic in the sense of Solonnikov allows to apply to them some well-posedness results available for initial-boundary value problems for the Solonnikov parabolic systems, see, e.g., [7], [5, Chapter VII, Section 10], [2, Section VI.3.6], [9].

References

1. B.R. Duffy, Flow of a liquid with an anisotropic viscosity tensor. J. Nonnewton. Fluid Mech. **4**, 177–193 (1978)
2. S.D. Eidelman, N.V. Zhitarashu: *Parabolic Boundary Value Problems* (Birkhäuser, Basel, 1998)
3. M. Kohr, S.E. Mikhailov, W.L. Wendland, Dirichlet and transmission problems for anisotropic Stokes and Navier-Stokes systems with L_∞ tensor coefficient under relaxed ellipticity condition. Discrete Contin. Dyn. Syst. Ser. A. **41**, 4421–4460 (2021)
4. M. Kohr, S.E. Mikhailov, W.L. Wendland, Layer potential theory for the anisotropic Stokes system with variable L_∞ symmetrically elliptic tensor coefficient. Math. Methods Appl. Sci. **44**, 9641–9674 (2021)
5. O.A. Ladyzhenskaya, V.A. Solonnikov, N.N. Ural'tseva, *Linear and Quasilinear Equations of Parabolic Type* (Nauka, Moscow, 1967); English transl.: Amer. Math. Soc., Providence, RI, 1968
6. O.A. Oleinik, A.S. Shamaev, G.A. Yosifian, *Mathematical Problems in Elasticity and Homogenization*, (Horth-Holland, Amsterdam, 1992)
7. V.A. Solonnikov, On boundary value problems for linear parabolic systems of differential equations of general form. Trudy Mat. Inst. Steklov. **83**, 3–163 (1965); English transl. in Proc. Steklov Inst. Math. **83** (1965). [MR 35 #1965]
8. R. Temam, *Navier-Stokes Equations. Theory and Numerical Analysis*, AMS Chelsea Edition (American Mathematical Society, 2001)
9. N.V. Zhitarashu, L_2-theory of generalized solutions of general linear parabolic boundary value problems. Mat. Issled. **112**, 104–115 (1990) (in Russian). [MR 91j:35129]

On Convergence and Divergence of Fourier Series and Féjer Means with Applications to Lebesgue and Vilenkin-Lebesgue Points

Nato Nadirashvili and Tural Nasibov

Abstract In the first part of this paper review results concerning almost everywhere convergence and divergence on the set of measure zero of partial sums and Féjer means with respect to Vilenkin and trigonometric systems. By using this information we prove that for any set E of measure zero there exists a function $f \in L_1(G_m)$ such that each $x \in E$ is not Lebesgue and Vilenkin-Lebesgue points.

Keywords Fourier analysis · Vilenkin Group · Vilenkin system · Fourier series · Partial sums · Féjer means · Almost everywhere convergence · Divergence on sets of measure zero · Carleson-Hunt theorem · Kolmogorov theorem

2000 Mathematics Subject Classification 42C10, 42B25

1 Vilenkin Groups and Functions

Denote by $\mathbb{N}_+$ the set of the positive integers, $\mathbb{N} := \mathbb{N}_+ \cup \{0\}$. Let $m := (m_0, m_1, \ldots)$ be a sequence of the positive integers not less than 2. Denote by

$$Z_{m_k} := \{0, 1, \ldots, m_k - 1\}$$

the additive group of integers modulo m_k. Define the group G_m as the complete direct product of the groups Z_{m_i} with the product of the discrete

The research was supported by Shota Rustaveli National Science Foundation (SRNSF) grant no. MR-23-173.

N. Nadirashvili (✉) · T. Nasibov
School of Science and Technology, The University of Georgia, Tbilisi, Georgia
e-mail: n.nadirashvili@ug.edu.ge

R. Duduchava et al. (eds.), *Tbilisi Analysis and PDE Seminar*, Research Perspectives Ghent Analysis and PDE Center 7,
https://doi.org/10.1007/978-3-031-62894-8_16

topologies of Z_{m_j}'s. The elements of G_m are represented by sequences $x := (x_0, x_1, \ldots, x_j, \ldots)$, $(x_j \in Z_{m_j})$.

The direct product μ of the measures $\mu_k(\{j\}) := 1/m_k \ (j \in Z_{m_k})$ is the Haar measure on G_m with $\mu(G_m) = 1$.

If we define the so-called generalized number system based on m by

$$M_0 := 1, \ M_{k+1} := m_k M_k \quad (k \in \mathbb{N}),$$

then every $n \in \mathbb{N}$ can be uniquely expressed as $n = \sum_{j=0}^{\infty} n_j M_j$, where $n_j \in Z_{m_j}$ $(j \in \mathbb{N}_+)$ and only a finite number of n_j's differ from zero.

First, we define the generalized Rademacher functions $r_k(x) : G_m \to \mathbb{C}$ by

$$r_k(x) := \exp(2\pi \imath x_k/m_k), \ \left(\imath^2 = -1, \ x \in G_m, \ \ k \in \mathbb{N}\right).$$

Now, define the Vilenkin system $\psi := (\psi_n : n \in \mathbb{N})$ on G_m by (see [47, 53, 54]):

$$\psi_n(x) := \prod_{k=0}^{\infty} r_k^{n_k}(x), \quad (n \in \mathbb{N}).$$

Specifically, we call this system the Walsh-Paley system when $m \equiv 2$.

The Vilenkin system is orthonormal and complete in $L_2(G_m)$ (see e.g. [1]).

2 Partial Sums and Féjer Means with Respect to Vilenkin Systems

For $f \in L_1(G_m)$, we define the Fourier coefficients, the partial sums with respect to Vilenkin systems, Féjer means with respect to Vilenkin and systems in the usual manner:

$$\widehat{f}(n) := \int_{G_m} f \overline{\psi}_n d\mu, \quad (n \in \mathbb{N}),$$

$$S_n f := \sum_{k=0}^{n-1} \widehat{f}(k)\, \psi_k, \quad (n \in \mathbb{N}_+)$$

and

$$\sigma_n f := \frac{1}{n} \sum_{k=0}^{n-1} S_k f, \quad (n \in \mathbb{N}_+)$$

respectively.

We also define the maximal operators S^* and σ^* of partial sums and Féjer means with respect to Vilenkin and trigonometric systems in the following way:

$$S^* f := \sup_{n\in\mathbb{N}} |S_n f| \quad \text{and} \quad \sigma^* f := \sup_{n\in\mathbb{N}} |\sigma_n f|.$$

The history of the Fourier series began with the problem to solve some partial differential equations in the middle of the eighteenth century after preliminary investigations by Euler, d'Alembert, and Bernoulli. In 1822 Fourier published his famous book [15], where he used the trigonometric series to write down the solution of the heat equation with boundary conditions. After this useful application, the problem of convergence of the Fourier series arose. The first theorem related to the convergence of the Fourier series was proved by Dirichlet in 1829. He determined sufficient condition which guarantees convergence at a given point. After this theorem Lipschitz, Jordan, and Dini, either generalized Dirichlet's results or found other criteria for the convergence. Later on, the majority shared the idea that the Fourier series of a continuous function had to converge at any point. However, this was proved to be false by du Bois-Reymond [10] in 1873, who showed that there exists a continuous function whose Fourier series diverges at a point. Since then several other such constructions have been found. We mention Fejer's example [58] which is remarkable for its elegance and simplicity.

Almost-everywhere convergence of Fourier series for L_2 functions was postulated by Luzin [30] in 1915 and the problem was known as Luzin's conjecture. Carleson's theorem is a fundamental result in mathematical analysis establishing the pointwise (Lebesgue) almost everywhere convergence of Fourier series of L_2 functions, proved by Carleson [12] in 1966. The name of Carleson is also often used to refer to the extension of the result by Hunt [21], which was given in 1968 to L_p functions for $p \in (1, \infty)$ (also known as the Carleson-Hunt theorem). Carleson's original proof is exceptionally hard to read, and although several authors have simplified the arguments there are still no easy proof of his theorem. Expositions of the original Carleson's paper were published by Kahane [23], Mozzochi [31], Jorsboe and Mejlbro [22] and Arias de Reyna [39]. Moreover, Fefferman [14] published a new proof of Hunt's extension, which was done by bounding a maximal operator of partial sums $S^* f := \sup_{n\in\mathbb{N}} |S_n f|$. This, in turn, inspired a much simplified proof of the L_2 result by Lacey and Thiele [29], explained in more detail in Lacey [28]. In the books Fremlin [16] and Grafakos [19] it was also given proofs of Carleson's theorem. Already in 1923, Kolmogorov [27] showed that the analogue of Carleson's result for L_1 is false by finding a function whose Fourier series diverges almost everywhere (improved slightly in 1926 to diverging everywhere). The analogue of Carleson's theorem for Walsh system was proved by Billard [6] for $p = 2$ and by Sjölin [45] for $1 < p < \infty$, while for bounded Vilenkin systems it was proved by Gosselin [18]. Schipp [42] investigated the so called tree martingales, i.e., martingales with respect to a stochastic basis indexed by a tree, and generalized the results about maximal function, quadratic variation and martingale transforms to these martingales (see also [43, 57]). Using these results, he gave a proof of

Carleson's theorem for Walsh-Fourier series. A similar proof for bounded Vilenkin systems can be found in Schipp and Weisz [43, 57].

Recent proof of almost everywhere convergence of Walsh-Fourier series was given by Demeter [13] in 2015 and of Vilenkin-Fourier series was proved in [37] and [38] in 2022. By using some methods of martingale Hardy spaces, almost everywhere convergence of subsequences of Vilenkin-Fourier series was considered in [9] (see also [4, 5, 8, 36, 48, 49, 51, 52, 55]). It follows that, if we consider the restricted maximal operator $\widetilde{S}_{\#}^{*}$, defined by

$$\widetilde{S}_{\#}^{*} f := \sup_{n\in\mathbb{N}} \left|S_{M_n} f\right|, \quad (M_k := m_0 \ldots m_{k-1}, \quad k = 0, 1 \ldots), \tag{2.1}$$

we get that

$$y\mu \left\{\widetilde{S}_{\#}^{*} f > y\right\} \leq c \left\|f\right\|_1, \quad f \in L^1(G_m), \quad y > 0.$$

Hence, if $f \in L_1(G_m)$, then $S_{M_n} f \to f,$ a.e. on G_m.

In particular, we can characterize such points where convergence holds. Thus, we need the following definition:

Definition 2.1 In the one-dimensional case a point x is called a Lebesgue point of the integrable function f if

$$\lim_{h\to 0} \frac{1}{h} \int_x^{x+h} |f(t) - f(x)| \, d\mu(t) = 0.$$

It is known (see [37]) that the following is true:

Theorem 2.2 *Let $f \in L_1(G_m)$. Then the following two statements hold:*

(a) almost everywhere x of function f are Lebesgue points.
(b) For all Lebesgue points of f, we have that

$$\lim_{n\to\infty} S_{M_n} f(x) = f(x).$$

Stein [46] constructed an integrable function whose Walsh-Fourier series diverges almost everywhere. Later Schipp [40, 44] proved that there exists an integrable function whose Walsh-Fourier series diverges everywhere. Kheladze [25, 26] proved that for any set of measure zero there exists a function in $f \in L_p(G_m)$ $(1 < p < \infty)$ whose Vilenkin-Fourier series diverges on that set, while the result for continuous or bounded function was proved by Harris [20] or Bitsadze [7]. A resent proof of this result was given in [37]:

Theorem 2.3 *If $E \subseteq G_m$ is a set of measure zero, then there exists a function $f \in L_1(G_m)$ such that S_{M_n} partial sums with respect to Vilenkin systems diverge on this set.*

In the one-dimensional case the weak type inequality for the maximal operator σ^* of Fejér means

$$\mu\left(\sigma^* f > \lambda\right) \leq \frac{c}{\lambda} \|f\|_1 \qquad (\lambda > 0)$$

can be found in Zygmund [58] for the trigonometric series, in Schipp [41] for Walsh series and in Pál, Simon [34] for bounded Vilenkin series (see also [3, 32, 33, 35, 50]).

It is known (see [58]) that the Fejér means $\sigma_n Tf$ of the trigonometric Fourier series of $f \in L_1$ converge to f at each Lebesgue point.

Goginava and Gogoladze defined Vilenkin-Lebesgue points (see [17]). A similar definition of Walsh-Lebesgue points is due to Weisz [56].

Definition 2.4 The operator W_A defined by

$$W_A f(x) := \sum_{s=0}^{A-1} M_s \sum_{r_s=1}^{m_s-1} \int_{I_A(x-r_s e_s)} |f(t) - f(x)|\, d\mu(t).$$

A point $x \in G_m$ is called a Vilenkin-Lebesgue point of a function $f \in L_1(G_m)$ if

$$\lim_{A\to\infty} W_A f(x) = 0.$$

In particular, Goginava and Gogoladze [17] proved the following theorem:

Theorem 2.5 *Let $f \in L_1(G_m)$. Then the following two statements hold:*

(a) almost everywhere x of function f are Lebesgue points.
(b) For all Vilenkin-Lebesgue points of f, it yields that

$$\lim_{n\to\infty} \sigma_n f(x) = f(x).$$

Bugadze [11] proved that for any set of measure zero there exists a function in $f \in L_1(G_m)$ whose Féjer means of Walsh-Fourier series diverges on the set, while the result for the Vilenkin system was proved by G. A. Karagulyan [24].

In particular, Karagulyan considered sequences of linear operators $U_n f(x)$ with some localization property, which is fulfilled for all summability methods with respect to orthonormal systems and proved that for any set E of measure zero there exists a set A for which $U_n I_A(x)$ diverges at each point $x \in E$. This result is a generalization of analogous theorems known for the Fourier sums operators with respect to different orthogonal systems. After some modification of the integrable function given in [38] whose Vilenkin-Fourier series diverges on set of measure zero, Areshidze and Tephnadze [2] constructed a function in $f \in L_1(G_m)$ whose Féjer means with respect to Vilenkin system diverges on the set of measure zero:

Theorem 2.6 *If $E \subseteq G_m$ is a set of measure zero, then there exists a function $f \in L_1(G_m)$ such that the Féjer means with respect to Vilenkin systems diverge on this set.*

3 Lebesgue and Vilenkin-Lebesgue Points of Integrable Functions

By using information above in this section we prove that for any set E of measure zero there exists a function $f \in L_1(G_m)$ such that each $x \in E$ is not Lebesgue and Vilenkin-Lebesgue points.

Theorem 3.1 *If $E \subseteq G_m$ is a set of measure zero, then there exists a function $f \in L_1(G_m)$ such that each $x \in E$ is not a Lebesgue point.*

Proof As it was mentioned in Theorem 2.2, that the S_{M_n}-th partial sums converge at each Lebesgue point. On the other hand, Theorem 2.6 says that for any set $E \subseteq G_m$ of measure zero, there exists a function $f \in L_1(G_m)$ such that the S_{M_n} partial sums with respect to Vilenkin systems diverge on this set, which means that they can not be Lebesgue points. The proof is complete. □

Theorem 3.2 *If $E \subseteq G_m$ is a set of measure zero, then there exists a function $f \in L_1(G_m)$ such that each $x \in E$ is not a Vilenkin-Lebesgue point.*

Proof As it was mentioned in Theorem 2.5, Féjer means with respect to Vilenkin systems converge at each Vilenkin-Lebesgue point. On the other hand, Theorem 2.3 says that for any set $E \subseteq G_m$ of measure zero, there exists a function $f \in L_1(G_m)$ such that Féjer means with respect to Vilenkin systems diverge on this set, which means that they can not be Vilenkin-Lebesgue points. The proof is complete. □

Acknowledgments I would like to thank professor Lars-Erik Persson for his very good suggestions, which have improved the paper. The research was supported by The University of Georgia grant "Mathematical Analysis and PDE on Lie Groups".

References

1. G.N. Agaev, N.Y. Vilenkin, G.M. Dzhafarly, A.I. Rubinshtein, *Multiplicative Systems of Functions and Harmonic Analysis on Zero-dimensional Groups* (Russian) ("Elm", Baku, 1981)
2. N. Areshidze, L.E. Persson, G. Tephnadze, A note on the divergence of Fejér means with respect to Vilenkin systems on sets of measure zero. Arab. J. Math. (to appear)
3. D. Baramidze, N. Nadirashvili, L.-E. Persson, G. Tephnadze, Some weak-type inequalities and almost everywhere convergence of Vilenkin-Nörlund means. J. Inequal. Appl., paper no. **66**, 17 pp. (2023)
4. D. Baramidze, L.-E. Persson, G. Tephnadze, Some new $(H_p - L_p)$ type inequalities for weighted maximal operators of partial sums of Walsh-Fourier series. Positivity **27**(3), paper no. 38, 14 pp. (2023)

5. D. Baramidze, I. Blahota, G. Tephnadze, R. Toledo, Martingale Hardy spaces and some new weighted maximal operator of Fejér means of Walsh-Fourier series. J. Geom. Anal. **34**, paper no. 3, 17 pp. (2024)
6. P. Billard, Sur la convergence presque partout des séries de Fourier-Walsh des fonctions de l'espace $L^2[0, 1]$. Studia Math. **28**, 363–388 (1966/1967)
7. K. Bitsadze, On divergence of Fourier series with respect to multiplicative systems on the sets of measure zero. Georgian Math. J. **16**(3), 435–448 (2009)
8. I. Blahota, L.E. Persson, G. Tephnadze, Two-sided estimates of the Lebesgue constants with respect to Vilenkin systems and applications. Glasg. Math. J. **60**(1), 17–34 (2018)
9. I. Blahota, K. Nagy, L.E. Persson, G. Tephnadze, A sharp boundedness result concerning maximal operators of Vilenkin-Fourier series on martingale Hardy spaces. Georgian Math. J. **26**(3), 351–360 (2019)
10. P.du. Bois-Reymond, Üeber die fourierschen reihen, Nachr. Kön. Ges. Wiss. Göttingen **21**, 571–582 (1873)
11. V.M. Bugadze, On divergence of Fourier–Walsh series of bounded functions on sets of measure zero. Russian Acad. Sci. Sb. Math. **82**(2), 365–372 (1995)
12. L. Carleson, On convergence and growth of partial sums of Fourier series. Acta Math. **116**, 135–157 (1966)
13. C. Demeter, A guide to Carleson's theorem. Rocky Mountain J. Math. **45**(1), 169–212 (2015)
14. Ch. Fefferman, Pointwise convergence of Fourier series. Ann. Math. **98**(3), 551–571 (1973)
15. J. Fourier, *Théorie analytique de la chaleur* (Gauthier-Villars et fils, 1888)
16. D.H. Fremlin, *Measure Theory. vol. 2, Broad Foundations*. Corrected second printing of the 2001 original (Torres Fremlin, Colchester, 2003)
17. U. Goginava, L. Gogoladze, Pointwise summability of Vilenkin-Fourier series. Publ. Math. Debrecen **79**(1–2), 89–108 (2011)
18. J.A. Gosselin, Almost everywhere convergence of Vilenkin-Fourier series. Trans. Amer. Math. Soc. **185**, 345–370 (1974)
19. L. Grafakos, *Classical Fourier Analysis*, 3rd edn., vol. 249. Graduate Texts in Mathematics (Springer, New York, 2014)
20. D.C. Harris, Compact sets of divergence for continuous functions on a Vilenkin group. Proc. Amer. Math. Soc. **98**(3), 436–440 (1986)
21. R. Hunt, On the convergence of Fourier series, in *Orthogonal Expansions and their Continuous Analogues, (Proc. Conf., Edwardsville, Ill., 1967)* (Southern Illinois University Press, Carbondale, 1968), pp. 235–255
22. O.G. Jorsboe, L. Mejlbro, *The Carleson-Hunt Theorem on Fourier Series*, vol. 911, Lecture Notes in Mathematics (Springer, Berlin, 1982)
23. J.P. Kahane, *Sommes partielles des séries de Fourier*, d'apres L. Carleson', Séminaire Bourbaki, 9 (Société Mathématique de France, Paris, 1995), pp. 491–507
24. G.A. Karagulyan, Divergence of general operators on sets of measure zero. Colloquium Math. **121**(1), 113–119 (2010)
25. Sh.V. Kheladze, On the everywhere divergence of Fourier-Walsh series. Sakharth. SSR Mecn. Acad. Moambe **77**, 305–307 (1975)
26. Sh.V. Kheladze, Divergence everywhere of Fourier series in bounded Vilenkin systems (Russian). Questions of the theory of functions and mathematical physics. Sakharth. SSR Mecn. Acad. Math. Inst. Strom. **58**, 224–242 (1978)
27. A.N. Kolmogorov, Une série de Fourier-Lebesgue divergente presque partout. Polska Akademii Nauk. Fund. Math. **4**, 324–328 (1923)
28. M.T. Lacey, Carleson's theorem: proof, complements, variations. Publ. Mat. **48**(2), 251–307 (2004)
29. M. Lacey, C. Thiele, A proof of boundedness of the Carleson operator. Math. Res. Lett. **7**(4), 361–370 (2000)
30. N.N. Luzin, *Collected Works. Vol. I. Metric Theory of Function and Functions of a Complex Variable* (Russian) (Izdat. Acad. Nauk. SSSR, Moscow, 1953)

31. C.J. Mozzochi, *On the Pointwise Convergence of Fourier Series*, Lecture Notes in Mathematics, vol. 199 (Springer, Berlin, 1971)
32. N. Nadirashvili, L.-E. Persson, G. Tephnadze, F. Weisz, Vilenkin-Lebesgue points and almost everywhere convergence of Vilenkin-Fejér means and applications. Mediterr. J. Math. **19**, 5, paper no. 239, 16 pp. (2022)
33. N. Nadirashvili, G. Tephnadze, G. Tutberidze, Almost everywhere and norm convergence of approximate identity and Fejér means of trigonometric and Vilenkin systems. Trans. A. Razmadze Math. Inst. **177**(3), 453–463 (2023)
34. J. Pal, P. Simon, On a generalization of the concept of derivate. Acta Math. Hung. **29**, 155–164 (1977)
35. L.E. Persson, G. Tephnadze, P. Wall, On the maximal operators of Vilenkin-Nórlund means. J. Fourier Anal. Appl. **21**(1), 76–94 (2015)
36. L.E. Persson, G. Tephnadze, P. Wall, On an approximation of 2-dimensional Walsh-Fourier series in the martingale Hardy spaces. Ann. Funct. Anal. **9**(1), 137–150 (2018)
37. L.-E. Persson, G. Tephnadze, F. Weisz, *Martingale Hardy Spaces and Summability of One-dimensional Vilenkin-Fourier Series* (Birkhäuser/Springer, 2022)
38. L.-E. Persson, F. Schipp, G. Tephnadze, F. Weisz, An analogy of the Carleson-Hunt theorem with respect to Vilenkin systems. J. Fourier Anal. Appl. **28**, 48, 1–29 (2022)
39. J.A. Reyna, *Pointwise Convergence of Fourier Series*, Lecture Notes in Mathematics, vol. 1785 (Springer, Berlin, 2002)
40. F. Schipp, Über die divergenz der Walsh-Fourierreihen. Ann. Univ. Sci. Budapest Eötvös Sect. Math. **12**, 49–62 (1969)
41. F. Schipp, Certain rearrangements of series in the Walsh series. Mat. Zametki **18**, 193–201 (1975)
42. F. Schipp, Pointwise convergence of expansions with respect to certain product systems. Anal. Math. **2**(1), 65–76 (1976)
43. F. Schipp, F. Weisz, Tree martingales and almost everywhere convergence of Vilenkin-Fourier series. Math. Pannon. **8**(1), 17–35 (1997)
44. F. Schipp, W.R. Wade, P. Simon, J. Pál, *Walsh Series. An Introduction to Dyadic Harmonic Analysis* (Adam Hilger, Bristol, 1990)
45. P. Sjölin, An inequality of Paley and convergence a.e. of Walsh-Fourier series. Ark. Mat. **7**, 551–570 (1969)
46. E.M. Stein, On limits of sequences of operators. Ann. Math. **74**(2), 140–170 (1961)
47. N.Y. Vilenkin, On a class of complete orthonormal systems. Amer. Math. Soc. Trans. **28**(2), 1–35 (1963)
48. G. Tephnadze, On the partial sums of Vilenkin-Fourier series. J. Contemp. Math. Anal. **49**(1), 23–32 (2014)
49. G. Tephnadze, On the partial sums of Walsh-Fourier series. Colloq. Math. **141**(2), 227–242 (2015)
50. G. Tephnadze, Martingale Hardy spaces and summability of the one dimensional Vilenkin-Fourier series, Ph.D. thesis, Department of Engineering Sciences and Mathematics, Luleå University of Technology, Oct. 2015 (ISSN 1402–1544)
51. G. Tephnadze, On the convergence of partial sums with respect to Vilenkin system on the martingale Hardy spaces. J. Contemp. Math. Anal. **53**(5), 294–306 (2018)
52. G. Tephnadze, The one-dimensional martingale Hardy spaces and partial sums and Fejér means with respect to Walsh system. Mem. Differential Equations Math. Phys. **88**, 109–158 (2023)
53. N.Y. Vilenkin, On the theory of lacunary orthogonal systems (Russian). Izv. Akad. Nauk SSSR Ser. Mat. **13**, 245–252 (1949)
54. N.Y. Vilenkin, On the theory of Fourier integrals on topological groups. Mat. Sbornik N. S. **30**(72), 233–244 (1952)
55. C. Watari, On generalized Walsh-Fourier series. Tóhoku Math. J. **10**(2), 211–241 (1958)

56. F. Weisz, Convergence of singular integrals. Ann. Univ. Sci. Budapest Sect. Math. **32**, 243–256 (1989)
57. F. Weisz, *Martingale Hardy Spaces and their Applications in Fourier Analysis*. Lecture Notes in Mathematics, vol. 1568 (Springer, Berlin, 1994)
58. A. Zygmund, Smooth functions. Duke Math. J. **12**, 47–76 (1945)

On Generalized Sharpness of Some Hardy-Type Inequalities

Lars-Erik Persson, Natasha Samko, and George Tephnadze

Abstract The current status concerning Hardy-type inequalities with sharp constants is presented in a unified convexity way. In particular, it is then natural to replace the Lebesgue measure dx with the Haar measure dx/x. We also present some new two-sided Hardy-type inequalities for monotone functions, where not only the two constants are sharp but also where the involved function spaces are optimal. As applications, a number of both well-known and new Hardy-type inequalities are pointed out. These results are used to present some new information concerning sharpness in the relation between different quasi-norms in Lorentz spaces.

Keywords Inequalities · Hardy-type inequalities · Sharp constants · Optimal target function · Lorentz spaces

2010 Mathematics Subject Classification 26D10, 46E30

The research was supported by Shota Rustaveli National Science Foundation (SRNSF) grant no. FR-21-2844, Project title: "Moving Manifolds".

L.-E. Persson · N. Samko (✉)
Department of Computer Science and Computational Engineering, UiT - The Arctic University of Norway, Narvik, Norway

G. Tephnadze
School of Science and Technology, The University of Georgia, Tbilisi, Georgia
e-mail: g.tephnadze@ug.edu.ge

R. Duduchava et al. (eds.), *Tbilisi Analysis and PDE Seminar*,
Research Perspectives Ghent Analysis and PDE Center 7,
https://doi.org/10.1007/978-3-031-62894-8_17

1 Introduction

The continuous Hardy inequality from 1925 (see [2]) informs us if f is non-negative p-integrable function on $(0, \infty)$, then f is integrable over the interval $(0, x)$ for each positive x and

$$\int_0^\infty \left(\frac{1}{x}\int_0^x f(y)dy\right)^p dx \leq \left(\frac{p}{p-1}\right)^p \int_0^\infty f^p(x)dx, \quad p > 1. \tag{1.1}$$

The development of the famous Hardy inequality in both discrete and continuous forms during the period 1906–1928 has its own history or, as we call it, prehistory. Contributions of mathematicians other than G.H. Hardy such as E. Landay, G. Polya, E. Schur and M. Reisz, are important here. This prehistory was described in [6].

The first weighted version of (1.1) was proved by Hardy himself in 1928 (see [3]):

$$\int_0^\infty \left(\frac{1}{x}\int_0^x f(y)dy\right)^p x^a dx \leq \left(\frac{p}{p-1-a}\right)^p \int_0^\infty f^p(x)x^a dx \tag{1.2}$$

where f is measurable and non-negative function on $(0, \infty)$ whenever $a < p - 1, p > 1$.

In the remarkable further development to which today is called Hardy-type inequalities, in the case of weighted Lebesgue spaces mostly the Lebesgue measure dx is used (see the books [4–7] and the references therein). One basic idea in this paper is to use convexity and then it is more natural to instead use the measure dx/x (=Haar measure when the underlying group is R_+). Moreover, this way to consider the situation helps us to easier investigate and describe the sharpness in Hardy-type inequalities. In this theory of Hardy-type inequalities (between weighted Lebesgue spaces) we usually have good estimates of the sharp constant (= the operator norm or quasi-norm). However, in very few cases the sharp constant is known.

In this paper we describe and/or derive most of these Hardy-type inequalities in the convexity (dx/x) frame described above. Moreover, we concentrate also on the problem to derive the corresponding reversed inequalities in cones of monotone functions. And still with sharp constants. It turns out that our approach also implies that the sharpness can be further improved in special situations e.g. to not only have sharp constant(s) but also by involving more optimal function spaces, sometimes even with optimal so called target functions involved. In order to illustrate this idea we present the following introductory example:

Example 1.1 The inequality (1.2) holds also if the interval $(0, \infty)$ is replaced by $(0, \ell)$, $0 < \ell \leq \infty$, and still the constant $C = \left(\frac{p}{p-1-\alpha}\right)^p$ is sharp. However, also the following "sharper" and more general inequality is known (see [9] and cf. Theorem 2.3 (a) in the book [7]):

$$\int_0^\ell \left(\frac{1}{x}\int_0^x f(y)dy\right)^p x^\alpha dx \leq \left(\frac{p}{p-1-a}\right)^p \int_0^\ell f^p(x)x^\alpha \left[1-\left(\frac{x}{\ell}\right)^{\frac{p-1-a}{p}}\right]dx, \tag{1.3}$$

where $\alpha < p-1$, $p > 1$, and still the constant $C = \left(\frac{p}{p-1-a}\right)^p$ is sharp. Moreover, we note that in the cone of non-increasing functions (1.2) holds in the reversed direction with the constant $C = 1$. But indeed (1.3) holds also in the reversed direction with the sharp constant $C = p/(p-1-a) > 1$ whenever $\alpha > -1$. See our Theorem 3.1 (a). In such a situation when both constants are sharp we say that the involved weight function $g(x) := 1-\left(\frac{x}{\ell}\right)^{\frac{p-1-a}{p}} < 1$, $x < \ell$, is the "optimal target function".

2 A Convexity Approach to Derive Sharp Power Weighted Hardy-Type Inequalities

The fact that the concept of convexity can be used to prove several inequalities, both classical and new ones, was of course known by Hardy himself. For example in the famous book [4] this concept and the more or less equivalent Jensen inequality was frequently used. Hence, it may be surprising that Hardy himself never discovered that also his famous inequality in both original (see (1.1)) and power weighted form (see e.g. (1.2)) follow more or less directly as described below. Concerning convexity and its applications e.g. to prove inequalities we refer to the recent book [8], the papers [9, 10], and the references therein. We note that for $p > 1$ and $f(x) = g(x^{1-1/p})x^{-1/p}$:

$$\int_0^\infty \left(\frac{1}{x}\int_0^x f(y)dy\right)^p dx \leq \left(\frac{p}{p-1}\right)^p \int_0^\infty f^p(x)dx,$$

$$\Leftrightarrow$$

$$\int_0^\infty \left(\frac{1}{x}\int_0^x g(y)dy\right)^p \frac{dx}{x} \leq 1\cdot \int_0^\infty g^p(x)\frac{dx}{x}. \tag{2.1}$$

This means that Hardy's inequality (1.1) is equivalent to (2.1) for $p > 1$ and, thus, that Hardy's inequality can be proved in the following simple way (see form (2.1)): By Jensen's inequality and Fubini's theorem we have that

$$\int_0^\infty \left(\frac{1}{x}\int_0^x g(y)dy\right)^p \frac{dx}{x} \leq \int_0^\infty \left(\frac{1}{x}\int_0^x g^p(y)dy\right)\frac{dx}{x} = \int_0^\infty g^p(y)\frac{dy}{y}. \tag{2.2}$$

By instead making the substitution $f(t) = g(t^{\frac{p-1-a}{p}})t^{-\frac{1+a}{p}}$ in (1.2) we see that also this inequality is equivalent to (2.1). These facts imply especially the following:

(a) Hardy's inequalities (1.1) and (1.2) hold also for $p < 0$ (because the function $\varphi(u) = u^p$ is convex also for $p < 0$) and hold in the reverse direction for $0 < p < 1$ (with sharp constants $\left(\frac{p}{1-p}\right)^p$ and $\left(\frac{p}{a+1-p}\right)^p$, $a > p-1$, respectively).
(b) The inequalities (1.1) and (1.2) are equivalent, since both are equivalent to (2.1).
(c) The inequality (2.1) holds also with equality for $p = 1$, which gives us a possibility to interpolate and get more information about the mapping properties of the Hardy operator. In particular, we can use interpolation theory to see that in fact the Hardy operator H maps each interpolation space B between $L_1\left((0,\infty), \frac{dx}{x}\right)$ and $L_\infty\left((0,\infty), \frac{dx}{x}\right)$ into B, i.e. that the following more general Hardy type inequality holds: $\|Hf\|_B \le C\|f\|_B$.

For the finite interval case we need the following extension of our basic (convexity) form of Hardy's inequality presented above.

Theorem 2.1 *Let g be a non-negative and measurable function on $(0, \ell)$, $0 < \ell \le \infty$.*

(a) If $p < 0$ or $p \ge 1$, then

$$\int_0^\ell \left(\frac{1}{x}\int_0^x g(y)dy\right)^p \frac{dx}{x} \le 1 \cdot \int_0^\ell g^p(x)\left(1 - \frac{x}{\ell}\right)\frac{dx}{x}. \tag{2.3}$$

(In the case $p < 0$ we assume that $g(x) > 0$, $0 < x \le \ell$).
(b) If $0 < p \le 1$, then (2.2) holds in the reversed direction.
(c) The constant $C = 1$ is sharp in both (a) and (b).

By doing similar calculations as in the proof of Theorem 2.4 in [9] (see also Theorem 7.10 in the book [7]), or just doing appropriate transformations, we obtain the following symmetric version of this equivalence Theorem:

Theorem 2.2 *Let $0 < \ell \le \infty$, let $p \in \mathbb{R}_+ \setminus \{0\}$ and let f be a non-negative function. Then*

(a) the inequality

$$\int_0^\ell \left(\int_0^x f(y)dy\right)^p x^{-\alpha}\frac{dx}{x} \le \left(\frac{p}{\alpha}\right)^p \int_0^\ell (xf(x))^p x^{-\alpha}\left[1 - \left(\frac{x}{\ell}\right)^{\frac{\alpha}{p}}\right]\frac{dx}{x} \tag{2.4}$$

holds for all measurable functions f, each ℓ, $0 < \ell \le \infty$ and all α in the following cases:

(a_1) $p \ge 1, \alpha > 0$, (a_2) $p < 0, \alpha < 0$.

(b) *For the case* $0 < p < 1$, $\alpha > 0$, *inequality* (2.4) *holds in the reversed direction for all* ℓ, $0 < \ell \leq \infty$.

(c) *The inequality*

$$\int_{\ell}^{\infty}\left(\int_{x}^{\infty} f(y)dy\right)^{p} x^{\alpha}\frac{dx}{x} \leq \left(\frac{p}{\alpha}\right)^{p}\int_{\ell}^{\infty} f^{p}(x)x^{\alpha}\left[1-\left(\frac{\ell}{x}\right)^{\frac{\alpha}{p}}\right]\frac{dx}{x} \tag{2.5}$$

holds for all measurable functions f, *each* $\ell, 0 \leq \ell < \infty$ *and all* α *in the following cases:*

$$(c_1)\ \ p \geq 1, \alpha > 0, \qquad\qquad (c_2)\ \ p < 0, a < 0.$$

(d) *For the case* $0 < p \leq 1$, $\alpha > 0$ *inequality* (2.5) *holds in the reversed direction for all* ℓ, $0 \leq \ell < \infty$.

(e) *All inequalities above are sharp.*

(f) *Let* $p \geq 1$ *or* $p < 0$. *Then, the statements in (a) and (c) are equivalent for all permitted* α.

(g) *Let* $0 < p < 1$. *Then, the statements in (b) and (d) are equivalent for all permitted* α.

Remark 2.3 For the case $l = \infty$ the inequalities (2.4) and (2.5) were formulated, proved and applied in this convexity form in the new book [12]. This fact has further inspired us to reformulate our results in this convexity (dx/x) way, which not only contribute to a better understanding but is also more suitably for such applications in modern harmonic analysis.

3 Reversed Sharp Hardy Inequalities for Monotone Functions

The results in this section are inspired by the paper [1], where it is proved and applied that e.g. the following inequality holds with the sharp constant p :

$$\left(\int_{a}^{b} f(y)dy\right)^{p} \geq p\int_{a}^{b}(y-a)^{p-1}(f(y))^{p}dy, \tag{3.1}$$

whenever f is non-increasing on (a, b), $\infty < a < b \leq \infty$ and $p \geq 1$.

In our first result we consider the case when f is non-increasing and note that then such a reversed Hardy-type inequality has meaning only if $0 < \alpha < p$ (since if not the involved integrals diverges for all non-trivial functions f):

Theorem 3.1 *Let $p > 0$, $0 < \alpha < p$ and let f be a measurable, non-negative and non-increasing function on $(0, \ell)$, $0 < l \le \infty$.*

(a) *If $p \ge 1$, then*

$$\int_0^\ell \left(\int_0^x f(y)\,dy \right)^p x^{-\alpha} \frac{dx}{x} \ge \frac{p}{\alpha} \int_0^x (xf(x))^p x^{-\alpha} \left(1 - \left(\frac{x}{\ell}\right)^\alpha\right) \frac{dx}{x}. \tag{3.2}$$

(b) *If $0 < p \le 1$, then (3.2) holds in the reversed direction.*
(c) *The constant $C = p/\alpha$ is sharp in both (a) and (b) and equality appears for each function $f(x) = A\chi_{(0,c)}(x)$ for some $c \in (0, l)$ and $A > 0$.*

As already mentioned the inequality (3.2) has no meaning in the cone of non-increasing functions if $\alpha \ge p$. But it is not so if we instead restrict to the cone of non-decreasing functions. But in this case the "target function" $f(x) = \left(1 - \left(\frac{x}{\ell}\right)^\alpha\right)$ is different and connected to the truncated β_α function defined as follows:

$$\beta_\alpha = \beta_\alpha(u, v) = \int_\alpha^1 t^{u-1}(1-t)^{v-1}, \quad 0 \le \alpha < 1.$$

In particular β_0 coincides with usual β function $\beta(u, v)$. Our next main result reads:

Theorem 3.2 *Let $\alpha \ge p > 0$ and let f be a measurable, non-negative and non-decreasing function on $(0, \ell)$, $0 < \ell \le \infty$.*

(a) *If $p \ge 1$, then*

$$\int_0^\ell \left(\int_0^x f(y)\,dy \right)^p x^{-\alpha} \frac{dx}{x} \ge \frac{p}{\alpha} \int_0^\ell (xf(x))^p x^{-\alpha} T(x) \frac{dx}{x}, \tag{3.3}$$

where $T(x) := \alpha\beta_{\frac{x}{\ell}}(p, \alpha - p + 1)$, $x \le l$.
(b) *If $0 < p \le 1$, then (3.3) holds in the reversed direction.*
(c) *the constant $\frac{p}{\alpha}$ is sharp in both (a) and (b) and equality appears if*

$$f(x) = A\chi_{(c,l)}(x) \quad \text{for some} \quad c \in (0, l) \quad \text{and} \quad A > 0.$$

Example 3.3 For the case $l = \infty$ we obtain the sharp inequality

$$\int_0^\infty \left(\int_0^x f(y)\,dy \right)^p x^{-\alpha} \frac{dx}{x} \ge p\beta(p, \alpha - p + 1) \int_0^\infty (xf(x))^p x^{-\alpha} \frac{dx}{x}$$

for all non-decreasing functions f. This inequality holds in the reversed direction when $0 < p \leq 1$ and the constant is sharp also then. Hence, by just changing notations we see that our result generalizes also a result in [1].

Hence, we have investigated all cases concerning the usual (arithmetic mean) Hardy operator so we turn to the dual situation (cf. Theorem 2.2 c)) and here the only non-trivial situation is to study the non-increasing case.

Our main result for this case reads:

Theorem 3.4 *Let $p > 0$, $\alpha > 0$ and f be a measurable, non-negative and non-increasing function on (ℓ, ∞), $0 \leq \ell < \infty$.*

(a) If $p \geq 1$, then

$$\int_{\ell}^{\infty} \left(\int_{x}^{\infty} f(y)\, dy \right)^{p} x^{\alpha} \frac{dx}{x} \geq \frac{p}{\alpha} \int_{\ell}^{\infty} (xf(x))^{p} x^{\alpha} T_0(x) \frac{dx}{x}, \tag{3.4}$$

where $T_0(x) := \alpha \beta_{\frac{\ell}{x}}(p, \alpha)$, $x \geq l$.

(b) If $0 < p \leq 1$, then (3.4) holds in the reversed direction.

(c) The constant p/α is sharp in both (a) and (b) and equality appears in both (a) and (b) if $f(x) = A\chi_{(\ell,c)}(x)$ for some $c \in (\ell, \infty)$ and $A > 0$.

Example 3.5 Let f, p and α be defined as in Theorem 3.4. If $p \geq 1$, then

$$\int_{0}^{\infty} \left(\int_{x}^{\infty} f(y)\, dy \right)^{p} x^{\alpha} \frac{dx}{x} \geq p\beta(p, \alpha) \int_{0}^{\infty} (xf(x))^{p} x^{\alpha} \frac{dx}{x},$$

where $f(x)$ is a non-negative and non-increasing function. The inequality holds in the reversed direction when $0 < p \leq 1$ and the constant $p\beta(p, \alpha)$ is sharp in both cases. Hence, Theorem 3.4 may be regarded also as generalization of another result in [1].

4 Applications

By combining Theorem 2.2 (a), (b) and (c) with Theorem 3.1 we obtain the following sharp two sided estimates:

Theorem 4.1 *Let $p > 0$, $o < \alpha < p$, $0 < \ell \leq \infty$ and let f be a measurable, non-negative and non-increasing function on $(0, \ell)$. If $p > 1$, then*

$$\left(\frac{p}{\alpha}\right)^{1/p} I_1 \leq I_2 \leq \frac{p}{\alpha} I_1, \tag{4.1}$$

where

$$I_1 = \left(\int_0^{\ell} (xf(x))^p x^{-\alpha} \left(1 - \left(\frac{x}{\ell}\right)^{\alpha}\right) \frac{dx}{x} \right)^{1/p} \quad \text{and}$$

$$I_2 = \int_0^{\ell} \left(\int_0^{x} f(y)\, dy \right)^p x^{-\alpha} \frac{dx}{x}.$$

If $0 < p \leq 1$, *then* (4.1) *holds in the reversed direction. Moreover, both constants* $(p/\alpha)^{1/p}$ *and* p/α *are sharp for all* $p > 0$.

Remark 4.2

(*a*) This means that the equivalence $I_2 \approx I_1$ holds and the corresponding "optimal target function" is $g(x) = 1 - \left(\frac{x}{\ell}\right)^{\alpha}$.
(*b*) In the lower inequality we can even have equality while in the above inequality the sharpness follows by choosing a sequence of non-increasing functions (a well-known fact from the theory of Hardy-type inequalities).

Remark 4.3 Many crucial objects in different mathematical areas are non-decreasing (e.g. in Lorentz spaces, interpolation theory, approximation theory and harmonic analysis). Hence, in particular, Theorem 4.1 can be useful to obtain some more precise versions of known results in each of these areas. We illustrate this fact only in the theory of Lorentz spaces but aim to later also use our result to improve some results in the modern harmonic analysis as presented in the new book [12].

Let f^* denote the non-increasing rearrangement of a function f on a measure space (Ω, μ). The Lorentz spaces $L^{p,q}$, $0 < p, q < \infty$ are defined by using the quasi.norm (norm when $p > 1$, $q \geq 1$)

$$\|f\|^*_{p,q} := \left(\int_0^{\infty} \left(f^*(t)\, t^{1/p}\right)^q \frac{dt}{t} \right)^{1/q}. \tag{4.2}$$

It is well-known that for the case $p > 1$ this quasi-norm is equivalent to the following one equipped with the usual Hardy operator:

$$\|f\|^{**}_{p,q} := \left(\int_0^{\infty} \left(\int_0^{t} f^*(u)\, du \right)^q t^{-q/p'} \frac{dt}{t} \right)^{1/q}.$$

Moreover, we have the following more precise estimates:

$$\left(p'\right)^{1/q} \|f\|^*_{p,q} \leq \|f\|^{**}_{p,q} \leq p' \|f\|^*_{p,q} \tag{4.3}$$

if $q > 1$ and the reversed inequalities hold if $0 < q \leq 1$. However, by using Theorem 4.1 we not only get the sharp estimates in (4.3) but also the following more precise statement:

Corollary 4.4 *With the notations and assumptions above, $p > 1$ and $0 < \ell \leq \infty$, we have that*

$$\left(p'\right)^{1/q} I^*_\ell \leq I^{**}_\ell \leq p' I^*_\ell, \tag{4.4}$$

where $q > 1$,

$$I^*_\ell := \left(\int_0^\ell \left(f^*(t)\, t^{1/p}\right)^q \left(1 - \left(\frac{t}{\ell}\right)^{q/p'}\right) \frac{dt}{t}\right)^{1/q}$$

and

$$I^{**}_\ell := \left(\int_0^\ell \left(\int_0^t f^*(u)\, du\right)^q t^{-q/p'} \frac{dt}{t}\right)^{1/q}.$$

If $0 < q \leq 1$, then the inequalities in (4.4) *hold in the reversed directions. Both constants $(p')^{1/q}$ and p' are sharp for all $q > 0$.*

Remark 4.5 Note that (4.3) is obtained by just using (4.4) with $l = \infty$, so in particular, both constant in (4.3) (and the reversed inequalities for $0 \leq q \leq 1$) are sharp.

Remark 4.6 For the case $0 < p \leq 1$ it is known that the quasi-norm $\|f\|^*_{p,q}$ is equivalent to the following quasi-norm $\|f\|^{**}_{p,q}$ equipped with the dual Hardy operator:

$$\|f\|^{**}_{p,q} := \left(\int_0^\infty \left(\int_t^\infty f^*(u)\, du\right)^q t^{-q/p'} \frac{dt}{t}\right)^{1/q}$$

By instead using Theorem 2.2 *c*), *d*) and *e*) with $l = 0$ combined with Example 3.5 we obtain that if $0 < p \leq 1$, then

$$\left(q\beta\left(q, -q/p'\right)\right)^{1/q} \|f\|^*_{p,q} \leq \|f\|^{**}_{p,q} \leq -p' \|f\|^*_{p,q} \tag{4.5}$$

if $q \geq 1$ and the reversed inequalities hold if $0 < q \leq 1$. Both constants $\left(q\beta\left(q, -q/p'\right)\right)^{1/q}$ and $-p'$ are sharp for all $q > 0$.

Remark 4.7 A more general statement like that in Corollary 4.4 involving sharp constants in both inequalities can be formulated, where the integrals $\int\limits_0^\infty$ are replaced by the integrals $\int\limits_\ell^\infty$, $0 \leq \ell < \infty$. In particular, this gives a similar generalization of (4.5). However, in this case the result looks less nice since the two target functions $1 - \left(\frac{x}{\ell}\right)^\alpha$ and $\alpha\beta_{\frac{\ell}{x}}(p, \alpha)$ do not coincide.

We only give the following final example related to Remark 4.7 and the well-known inequality: If $0 < p < 1$, then

$$\int\limits_0^\infty \left(\frac{1}{x}\int\limits_x^\infty f(y)\,dy\right)^p dy \leq \frac{\pi p}{\sin \pi p}\int\limits_0^\infty f^p(x)\,dx \tag{4.6}$$

for all functions as defined in Theorem 3.4:

Example 4.8 Let $0 < p < 1$ and let f be a measurable, non-negative and non-increasing function on (ℓ, ∞), $0 \leq l < \infty$. If $0 < p \leq 1$, then

$$\int\limits_\ell^\infty \left(\frac{1}{x}\int\limits_x^\infty f(y)\,dy\right)^p dx \leq p\int\limits_\ell^\infty f^p(x)\,\beta_{\frac{\ell}{x}}(p, 1-p)\,dx,$$

and the constant p is sharp. This is just Theorem 3.4 (b) with $\alpha = 1 - p$. In particular, for $l = 0$ this inequality coincides with (4.6) since $\beta(p, 1-p) = \pi/\sin \pi p$, so the constant $\frac{\pi p}{\sin \pi p}$ in (4.6) is sharp.

Final Remark For detailed proofs at all results presented in this paper we refer [11]. In this paper sharpness is described also for the case $1 < p < q$, cf. also [10].

Acknowledgments The research was supported by The University of Georgia grant "Mathematical Analysis and PDE on Lie Groups."

References

1. J. Bergh, V. Burenkov, L.-E. Persson, Best constants in reversed Hardy's inequalities for quasimonotone functions. Acta Sci. Math. (Szeged) **59**, 221–239 (1994)
2. G.H. Hardy, Notes on some points in the integral calculus, LX. Messenger Math. **54**, 150–156 (1925)
3. G.H. Hardy, Notes on some points in the integral calculus, LXIV. Futher inequalities between integrals. Messenger Math. **57**, 12–16 (1928)

4. G.H. Hardy, J.E. Littlewood, G. Pólya, *Inequalities*, 2nd edn. (Cambridge University Press, 1952)
5. V. Kokilashvili, A. Meshki, L.-E Persson, *Weighted Norm Inequalities for Integral Transforms with Product Weights* (Nova Scientific Publishers, New York, 2010)
6. A. Kufner, L. Maligranda, L.-E. Persson, *The Hardy Inequality. About its History and Some Related Results, Vydavatelský Servis, Plzeň* (Vydavatelský Servis, Plzeň, 2007)
7. A. Kufner, L.-E. Persson, N. Samko, *Weighted Inequalities of Hardy Type*, 2nd edn. (Word Scientific, New Jersey, 2017)
8. C.P. Niculescu, L.-E. Persson, *Convex Functions and Their Applications. A Contemporary Approach*, 2nd edn., CMS Books in Mathematics (Springer, 2018)
9. L.-E. Persson, N. Samko, What should have happened if Hardy had discovered this? J. Inequal. Appl. **29**, 11 pp. (2012)
10. L.-E. Persson, S. Samko, A note on the best constants in some Hardy inequalities. J. Math. Inequal. **9**(2), 437–447 (2015)
11. L.-E. Persson, N. Samko, G. Tephnadze, Sharpness of some Hardy-type inequalities. J. Inequal. Appl. (2023), paper no. 155, 16 pp.
12. L.-E. Persson, G Tephnadze, F. Weisz, *Martingale Hardy Spaces and Summability of One-dimensional Vilenkin-Fourier Series* (Springer Nature, 2022)

Interaction Problems for n-Dimensional Dirac Operators with Singular Potentials

Vladimir Rabinovich

Abstract This paper is devoted to a brief review of the author's recent results for n-dimensional Dirac operators with singular potentials, supported on smooth bounded and unbounded hypersurfaces and some new results on the reduction of problems of interaction for Dirac operators with singular potentials to integral equations on hypersurfaces of interaction, and the connection between the Fredholm property of interaction problems and the mentioned integral operators.

Keywords Dirac operators · Singular potential · Delta-interactions · Self-adjointness · Essential spectrum

2000 Mathematics Subject Classification 2010 81Q10, 81V05, 35P15, 58C40

1 Introduction

The Dirac operator is one of the most important operators in physics and mathematics. Massive relativistic particles with spin (1/2), such as electrons and quarks, are described by three-dimensional Dirac operators. In dimension 2, the Dirac operators with special boundary conditions are used to describe the propagation of electrons in graphenes.

A large literature is devoted to boundary value problems for Dirac operators acting in sections of Clifford bundles, to their Fredholm properties, index formulas and applications to spectral geometry.

In the last decade, great interest has arisen in the study of Dirac operators with singular potentials in $\mathbb{R}^3$. These operators arise in the problems of confinement and transport of relativistic particles through surfaces which are supports of the

V. Rabinovich (✉)
Instituto Politecnico Nacional, ESIME Zac., Mexico City, Mexico

R. Duduchava et al. (eds.), *Tbilisi Analysis and PDE Seminar*,
Research Perspectives Ghent Analysis and PDE Center 7,
https://doi.org/10.1007/978-3-031-62894-8_18

confinement potentials (see, for instance, papers [1–16] and the literature cited there.)

This paper is devoted to a brief review of the author's recent results for n-dimensional Dirac operators with singular potentials, supported on smooth bounded and unbounded hypersurfaces (see [17, 18]) and some new ones results on the reduction of problems of interaction for Dirac operators with singular potentials to integral equations on hypersurfaces of interaction, and the connection between the Fredholm property of interaction problems and the mentioned integral operators.

The formal Dirac operators on $\mathbb{R}^n$ with singular potential is

$$D_{m,\Phi,\Gamma\delta_\Sigma} = \mathfrak{D}_{m,\Phi} + \Gamma\delta_\Sigma \tag{1}$$

where

$$\mathfrak{D}_{m,\Phi} = \alpha \cdot D_x + m\alpha_{n+1} + \Phi I_N = \sum_{j=1}^{n} \alpha_j \left(-i\partial_{x_j}\right) + m\alpha_{n+1} + \Phi I_N \tag{2}$$

is the Dirac operator on $\mathbb{R}^n$ with a regular electrostatic scalar potentials Φ, and with variable mass m of the particle, α_j, $j = 1, \ldots, n+1$ are the $N \times N$, $N = 2^{\left[\frac{n+1}{2}\right]}$ Dirac matrices, that is the Hermitian matrices satisfying the relations

$$\alpha_j\alpha_k + \alpha_k\alpha_j = 2\delta_{jk} I_N, \tag{3}$$

where I_N is the $N \times N$ unit matrix, $N = N(n) = 2^{[(n+1)/2]}$, $\Gamma\delta_\Sigma$ is the singular potential with the $N \times N$ strength matrix $\Gamma = \left(\Gamma_{i,j}\right)_{i,j=1}^{N}$ and δ_Σ is the delta-function with the support on a uniformly regular C^2-hypersurface $\Sigma \subset \mathbb{R}^n$ (see for instance [19]) being the common boundary of open sets $\Omega_\pm \subset \mathbb{R}^n$. We say that C^2-hypersurfaces $\Sigma \subset \mathbb{R}^n$ is **uniformly regular** if for fix $r > 0$ and for every point $x_0 \in \Sigma$ there exists a ball $B_r(x_0) = \{x \in \mathbb{R}^n : |x - x_0| < r\}$ and a diffeomorphism $\varphi_{x_0} : B_r(x_0) \to B_1(0)$ such that $\varphi_{x_0}(B_r(x_0) \cap \Sigma) = B_1(0) \cap \mathbb{R}^{n-1}_{y'}$, and

$$\sup_{x_0\in\Sigma}\ \sup_{|\alpha|\le l, x\in B_r(x_0)} \left|\partial^\alpha \varphi^i_{x_0}(x)\right| < \infty, i = 1, \ldots, n$$

where $\varphi^i_{x_0}$, $i = 1, \ldots, n$ are the coordinate functions of the mappings φ_{x_0}.

Let $H^1\left(\Omega_\pm, \mathbb{C}^N\right)$ be the Sobolev spaces of vector-valued distributions in domains $\Omega_\pm$, $H^1\left(\Omega\diagdown\Sigma, \mathbb{C}^N\right) = H^1\left(\Omega_+, \mathbb{C}^N\right) \oplus H^1\left(\Omega_-, \mathbb{C}^N\right)$. We associate with the formal Dirac operator $D_{\Phi,m,\Gamma\delta_\Sigma}$ the unbounded in $L^2(\mathbb{R}^n, \mathbb{C}^N)$ operator $\mathcal{D}_{\Phi,m,\mathfrak{B}_\Sigma}$ defined by the Dirac operator $\mathfrak{D}_{\Phi,m}$ with the domain

$$dom(\mathcal{D}_{m,\Phi,\mathfrak{B}_\Sigma}) = \left\{ \begin{array}{c} u \in H^1(\mathbb{R}^n\diagdown\Sigma, \mathbb{C}^N) : \\ \mathfrak{B}_\Sigma u(s) = a_+(s)\gamma^+_\Sigma u(s) + a_-(s)\gamma^-_\Sigma u(s) = 0, s \in \Sigma \end{array} \right\} \tag{4}$$

where $\gamma_\Sigma^\pm : H^1(\Omega_\pm \mathbb{C}^N) \to H^{1/2}(\Sigma, \mathbb{C}^N)$ are the trace operators,

$$a_\pm(s) = \frac{1}{2}\Gamma(s) \mp i\alpha \cdot \nu(s), s \in \Sigma, \alpha \cdot \nu(s) = \sum_{j=1}^{n} \alpha_j \nu_j(s), s \in \Sigma \tag{5}$$

$\nu(s), s \in \Sigma$ is the unit normal vectors to Σ directed into Ω_-.

We assume that $\Phi, m \in L^\infty(\mathbb{R}^n)$, and $\Gamma_{i,j} \in C_b^1(\Sigma)$, the space of differentiable functions on Σ bounded together with their derivatives.

2 Self-adjointness of the Unbounded Operators $\mathcal{D}_{\Phi,m,\mathfrak{B}_\Sigma}$

Theorem 1 *Let: (i) $\Phi, m \in L^\infty(\mathbb{R})$ be real-valued functions, (ii) Σ be a C^2-uniformly regular hypersurface, (iii) $\Gamma(x) \in C_b^1(\Sigma) \otimes \mathcal{B}(\mathbb{C}^N)$ be a Hermitian matrix for every $x \in \Sigma$, (iv) the uniform Lopatinsky-Shapiro condition*

$$\inf_{s\in\Sigma, \xi_s \in \mathbb{T}_s^*(\Sigma):|\xi_s|=1} \left| \det\left(\alpha \cdot \xi_s + \frac{\Gamma(s)}{2}\right)\right| > 0 \tag{6}$$

be satisfied, where $\mathbb{T}_s^*(\Sigma)$ is the cotangent space to Σ at the point $s \in \Sigma$. Then the unbounded operator $\mathcal{D}_{\Phi,m,,\mathfrak{B}_\Sigma}$ with domain

$$\left\{u \in H^1(\mathbb{R}^n \setminus \Sigma, \mathbb{C}^N) : \mathfrak{B}_\Sigma u(s) = a_+(s)\gamma_\Sigma^+ u(s) + a_-(s)\gamma_\Sigma^- u(s) = 0, s \in \Sigma\right\} \tag{7}$$

is self-adjoint in $L^2(\mathbb{R}^n, \mathbb{C}^N)$.

Example 2 Let the singular potential $\Gamma\delta_\Sigma$ be the sum of the singular magnetic,electrostatic, and the Lorentz potentials, that is

$$\Gamma(s) = \mathfrak{a}(s)\left(\alpha \cdot \nu(s)\right) + \mathfrak{b}(s)\mathbb{I}_N + \mathfrak{c}(s)\alpha_{n+1}, s \in \Sigma$$

where $\mathfrak{a}, \mathfrak{b}, \mathfrak{c} \in C_b^1(\Sigma)$ are real-valued functions. Then the condition

$$\inf_{s\in\Sigma} \left| 1 + \frac{\mathfrak{a}^2(s) + \mathfrak{c}^2(s) - \mathfrak{b}^2(s)}{4} \right| > 0. \tag{8}$$

implies condition (6). Therefore if Σ is a uniformly regular C^2-hypersurface, $\Phi, m \in L^\infty(\mathbb{R}^n)$ are real-valued functions, and condition (8) holds, then the operator $\mathcal{D}_{\Phi,m,\mathfrak{B}_\Sigma}$ is self-adjoint.

3 Fredholm Property of Interaction Problems

We also associate with the formal Dirac operator $D_{m,\Phi,\Gamma\delta_\Sigma}$ the operator $\mathbb{D}_{m,\Phi,\mathfrak{B}_\Sigma}$ of the interaction (transmission) problem

$$\mathbb{D}_{m,\Phi,\mathfrak{B}_\Sigma}u(x) = \begin{cases} \mathfrak{D}_{m,\Phi}u(x), x \in \mathbb{R}^n \diagdown \Sigma \\ \mathfrak{B}_\Sigma u(s), s \in \Sigma \end{cases}$$

acting from $H^1(\mathbb{R}^n \diagdown \Sigma, \mathbb{C}^N)$ into $L^2(\mathbb{R}^n, \mathbb{C}^N) \oplus H^{1/2}(\Sigma, \mathbb{C}^N)$.

We say that the function $a \in C_b^1(\mathbb{R}^n)$ is slowly oscillating at infinity and belongs to the class $SO(\mathbb{R}^n)$ if $\lim_{x\to\infty} \partial_{x_j} a(x) = 0$, $j = 1, \dots, n$.

Theorem 3 *Let $m, \Phi \in SO(\mathbb{R}^n)$, Σ be a closed C^2-hypersurface, $\Gamma \in C^1(\Gamma) \otimes \mathrm{hom}(\mathbb{C}^N)$. Then the operator $\mathbb{D}_{m,\Phi,\mathfrak{B}_\Sigma} : H^1(\mathbb{R}^n \diagdown \Sigma, \mathbb{C}^N) \to L^2(\mathbb{R}^n, \mathbb{C}^N) \oplus H^{1/2}(\Sigma, \mathbb{C}^N)$ is the Fredholm operator if:*

(i)

$$\liminf_{\mathbb{R}^n \ni x \to \infty} \mathrm{Re}(m^2(x) - \Phi^2(x)) > 0; \tag{9}$$

(ii)

$$\det\left(\alpha \cdot \xi_s + \frac{\Gamma(s)}{2}\right) \neq 0 \tag{10}$$

for each point $s \in \Sigma$ and $\xi_s \in \mathbb{T}_s^*(\Sigma)$.

Theorem 4 *Let $m, \Phi \in SO(\mathbb{R}^n)$, Σ be a C^2-hypersurface which is a conical set outside some ball $B_R(0) = \{x \in \mathbb{R}^n : |x| > R\}$, $\Gamma = (\Gamma_{kl})_{k.l=1}^N$, $\Gamma_{kl} \in SO(\Sigma)$. Then the operator $\mathbb{D}_{A,\Phi,m,\mathfrak{B}_\Sigma}$ is Fredholm if: (i) conditions (9), (10) hold, (ii)*

$$\liminf_{\Sigma \ni s \to \infty} \inf_{\xi \in \mathbb{T}_s^*(\Sigma)} \left| \det(\mathbb{I}_N + \frac{\Gamma(s)}{2} \frac{\alpha \cdot \xi + m(s)\alpha_{n+1} + \Phi(s)\mathbb{I}_N}{\sqrt{|\xi|^2 + m^2(s) - \Phi^2(s)}} \right| > 0. \tag{11}$$

4 Integral Equations Method for the Interaction Problems for Dirac Operators

We assume that $m, \Phi \in SO^2(\mathbb{R}^n) = C_b^2(\mathbb{R}^n) \cap SO(\mathbb{R}^n)$, and the Dirac operator $\mathfrak{D}_{m,\Phi}$ is invertible from $H^1(\mathbb{R}^n, \mathbb{C}^N)$ into $L^2(\mathbb{R}^n, \mathbb{C}^N)$. Then

$$\mathfrak{D}_{m,\Phi}^{-1} f(x) = \int_{\mathbb{R}^n} g_{m,\Phi}(x, y) f(y) dy, x \in \mathbb{R}^n$$

where $g_{m,\Phi} \in C^2((\mathbb{R}^n \times \mathbb{R}^n)\setminus\Theta) \otimes \hom(\mathbb{C}^N)$, Θ is the diagonal set in $\mathbb{R}^n \times \mathbb{R}^n$. Moreover,

$$\left\| g_{m,\Phi}(x,y) \right\|_{\hom(\mathbb{C}^N)} \leq C_{\alpha\beta} e^{-\varepsilon|x-y|}, \varepsilon > 0, |x-y| > 0, C_{\alpha\beta} > 0$$

for all α, β.

Let $\Sigma \subset \mathbb{R}^n$ be a C^2–uniformly regular hypersurface which is the common boundary of open domains $\Omega_\pm$.The potential operator on Σ associated with the invertible Dirac operator $\mathfrak{D}_{m,\Phi}$ is defined as follows

$$\begin{aligned} \mathcal{P}_{m,\Phi,\Sigma}\varphi(x) &= \mathfrak{D}^{-1}_{m,\Phi}\left(\varphi \otimes \delta_\Sigma\right)(x) \\ &= \int_\Sigma g_{m,\Phi}(x,s)\varphi(s)d\mu(s), \varphi \in C_0^\infty(\Sigma,\mathbb{C}^N), x \in \mathbb{R}^n\setminus\Sigma. \end{aligned} \quad (12)$$

where $d\mu$ is the surface Lebesgue measure on Σ.

Proposition 5 *Let $\Sigma \subset \mathbb{R}^n$ be a C^2–uniformly regular hypersurface. Then*

(i) $\mathcal{P}_{m,\Phi,\Sigma} : H^{1/2}(\Sigma,\mathbb{C}^N) \to H^1(\mathbb{R}^n\setminus\Sigma,\mathbb{C}^N)$ is a bounded operator;
(ii) There exists the non tangential limits

$$\begin{aligned} \mathcal{P}^\pm_{\Phi,m,\Sigma}\varphi(s) = \lim_{\Omega_\pm \ni x \to s \in \Sigma} \mathcal{P}_{\Phi,m,\Sigma}\varphi(s) &= \pm\frac{i}{2}\alpha \cdot \nu(s)\varphi(s) \\ &+ \mathcal{K}_{\Phi,m,\Sigma}\varphi(s), s \in \Sigma \end{aligned} \quad (13)$$

where $\nu(s)$ is the unit normal vector to Σ at the point s directed to Ω_-, $\mathcal{K}_{\Phi,m,\Sigma}$ is a singular integral operator with the symbol $\sigma\left(\mathcal{K}_{A,\Phi,m,\Sigma}\right) \in C_b^1(T^*(\Sigma),\mathbb{C}^N)$ where $T^*(\Sigma)$ is the cotangent bundle under Σ.
(iii) $\mathcal{P}^\pm_{\Phi,m,\Sigma}$ are bounded operators in $H^{1/2}(\Sigma,\mathbb{C}^N)$.

We will seek the solution of the interaction problem

$$\mathbb{D}_{m,\Phi,\mathfrak{B}_\Sigma}u(x) = \begin{cases} \mathfrak{D}_{\Phi,m}u(x) = f(x), x \in \mathbb{R}^n\setminus\Sigma \\ \mathfrak{B}_\Sigma u(s) = f_1(s), s \in \Sigma \end{cases} \quad (14)$$

of the form

$$u(x) = \mathfrak{D}^{-1}_{m,\Phi}f(x) + \mathcal{P}_{m,\Phi,\Sigma}\varphi(x), x \in \mathbb{R}^n\setminus\Sigma \quad (15)$$

where $\varphi \in H^{1/2}(\Sigma,\mathbb{C}^N)$. Substituting u defined by the formula (15) in the interaction condition $\mathfrak{B}_\Sigma u = f_1$ we obtain the singular integral equation on Σ

$$\Xi_{m,\Phi,\Sigma}\varphi(s) = (I + \Gamma\mathcal{K}_{\Phi,m,\Sigma})\varphi(s) = f_1(s) - \left(\mathfrak{B}_\Sigma\mathfrak{D}^{-1}_{\Phi,m}\right)f(s), s \in \Sigma.$$

In the case if Σ is a closed hypersurface we obtain the following result.

Theorem 6 *Assume that: (i)* $m.\Phi \in SO^2(\mathbb{R}^n)$*, (ii) the operator* $\mathfrak{D}_{m,\Phi} : H^1(\mathbb{R}^n, \mathbb{C}^N) \to L^2(\mathbb{R}^n, \mathbb{C}^N)$ *is invertible, (iii)* Σ *is a closed* C^2-*hypersurface, (iv)* $\Gamma \in SO(\Sigma) \otimes \mathrm{hom}(\mathbb{C}^N)$*. Then*

$$\mathbb{D}_{m,\Phi,\mathfrak{B}} : H^1(\mathbb{R}^n \diagdown \Sigma, \mathbb{C}^N) \to L^2(\mathbb{R}^n \diagdown \Sigma, \mathbb{C}^N) \oplus H^{1/2}(\Sigma, \mathbb{C}^N) \tag{16}$$

is a Fredholm operator if and only if the singular integral operator $\Xi_{m,\Phi,\Sigma} : H^{1/2}(\Sigma, \mathbb{C}^N) \to H^{1/2}(\Sigma, \mathbb{C}^N)$ *is a Fredholm operator. This condition is equivalent the condition (10).*

In the case of unbounded hypersurfaces the formulation of Theorem 6 is more complicated and it takes into account conditions at infinity.

Theorem 7 *We assume that: (i)* $m.\Phi \in SO^2(\mathbb{R}^n)$*,(ii) the operator* $\mathfrak{D}_{m,\Phi} : H^1(\mathbb{R}^n, \mathbb{C}^N) \to L^2(\mathbb{R}^n, \mathbb{C}^N)$ *is invertible; (iii)* Σ *is a* C^2-*conical hypersurface outside some ball;* $\Gamma \in SO(\Sigma) \otimes \mathrm{hom}\,\mathbb{C}^N$*. Then*

$$\mathbb{D}_{m,\Phi,\mathfrak{B}} : H^1(\mathbb{R}^n \diagdown \Sigma, \mathbb{C}^N) \to L^2(\mathbb{R}^n \diagdown \Sigma, \mathbb{C}^N) \oplus H^{1/2}(\Sigma, \mathbb{C}^N) \tag{17}$$

is a Fredholm operator if and only if the operator $\Xi_{m,\Phi,\Sigma} : H^{1/2}(\Sigma, \mathbb{C}^N) \to H^{1/2}(\Sigma, \mathbb{C}^N)$ *is a Fredholm operator. Necessary and sufficient conditions for* $\Xi_{m,\Phi,\Sigma}$ *to be a Fredholm operator are conditions (10) and (11).*

References

1. N. Arrizabalaga, A. Mas, L. Vega, Shell interactions for Dirac operators. J. Math. Pures Appl. (9) **102**(4), 617–639 (2014)
2. J. Behrndt, P. Exner, M. Holzmann, V. Lotoreichik, On the spectral properties of Dirac operators with electrostatic δ-shell interactions. J. Math.Pures Appl. **111**, 47–78 (2018)
3. J. Behrndt, P. Exner, M. Holzmann, V. Lotoreichik, On Dirac operators in $\mathbb{R}^3$ with electrostatic and Lorentz scalar δ-shell interactions. Quantum Stud. Math. Found. (2019). https://doi.org/10.1007/s40509-019-00186-6
4. J. Behrndt, M. Holzmann, T. Ourmières-Bonafos, K. Pankrashkin, Two-dimensional Dirac operators with singular interactions supported on closed curves. J. Funct. Anal. **279**(8), 108700 (2020)
5. J. Behrndt, M. Holzmann, Ch. Stelzer, G. Stenzel, Boundary triples and Weyl functions for Dirac operators with singular interactions (2022), arXiv:2211.05191 [math.SP]
6. J. Behrndt, M. Holzman, M. Tusek, Two-dimensional Dirac operators with general δ-shell interactions supported on a straight line (2022), arXiv:2208.12761v1 [math-ph]
7. B. Benhellal, Spectral Properties of the Dirac Operator coupled with $\delta-$Shell Interactions (2021), arXiv:2102.10207 [math.SP]
8. B. Benhellal, K. Pankrashkin, Curvature contribution to the essential spectrum of Dirac operators with critical shell interactions (2022). arXiv:2211.10264 [math.SP]
9. B. Cassano, V. Lotoreichik, A. Mas, M. Tusek, General δ–shell interactions for two-dimensional Dirac operators: self-adjointness and approximation (2021). arXiv:2102.09988v1 [math.AP]

10. M. Holzmann, A note on the three dimensional dirac operator with zigzag type boundary conditions. Complex Anal. Oper. Theory **15**, 47 (2021). https://doi.org/10.1007/s11785-021-01090-x
11. A. Mas, F. Pizzichillo, Klein's paradox and the relativistic $\delta-$shell interaction in $\mathbb{R}^3$. Anal. PDE **11**(3), 705–744 (2018)
12. J. Mehringer, E. Stockmeyer: Confinement–deconfinement transitions for two-dimensional Dirac particles. J. Funct. Anal. **266**, 2225–2250 (2014)
13. A. Moroianu, Ourmierès-Bonafos-Bonafos,Th.,Pankrashkin, K.: Dirac operators on surfaces large mass limits. J. Math. Pures Appl. **102**(4), 617–639 (2014)
14. F. Pizzichillo, H. Van Den Bosch, Self-adjointness of two-dimensional Dirac operators on corner domains. Preprint arXiv:1902.05010 (2019)
15. Th. Ourmierès-Bonafos-Bonafos, L. Vega, A strategy for self-adjointnessof Dirac operators: Applications to the MIT BAG model and shell interactions. Publ. Mat. **62**, 397–437 (2018)
16. Th. Ourmierès-Bonafos-Bonafos, F. Pizzichlllo, Dirac operators and shell interactions: a survey (2019), arXiv:1902.03901v1 [math-ph]
17. V. Rabinovich, Dirac operators with delta-Interactions on smooth hypersurfaces in $\mathbb{R}^n$. J. Fourier Anal. Appl. **28**, 20 (2022). https://doi.org/10.1007/s00041-022-09917-7
18. V. Rabinovich, Fredholm property of interaction problems on unbounded C^2-hypersurfaces in $\mathbb{R}^n$ for Dirac operators. J. Math. Sci. **271**, 136–161 (2023). https://doi.org/10.1007/s10958-023-06326-z
19. N. Grosse, V. Nistor, Uniform Shapiro-Lopatinski conditions and boundary value problems on manifolds with bounded geometry. Potential Anal. **53**, 407–447 (2020)

Convergence and Summability in Classical and Martingale Hardy Spaces

George Tephnadze

Abstract The development of the theory of martingale Hardy spaces has been strongly influenced by the classical theory. Because of this fact it is inevitable to compare results of this theory to those on classical Hardy spaces. Many similarities between these theories have already been found and a lot of new results were obtained in the theory of martingale Hardy spaces which are open problems in classical case. By taking this into account, in the paper we have chosen a partly opposite approach by stating and proving some new results concerning partial sums two-dimensional Fourier series in martingale Hardy spaces for the classical Hardy space case.

Keywords Walsh-Fourier series · Partial sums · Lebesgue space · Weak Lebesgue space · Martingale Hardy space · Maximal operators · Weighted maximal operators · Inequalities

2010 Mathematics Subject Classification 42C10

1 Introduction and Status of the Art

Let $\mathbb{N} := \{0, 1, \ldots\}$, $\mathbb{P} := \mathbb{N} \backslash \{0\}$ and $\mathbb{R}$ be the set of real numbers.

First, we define classical Hardy spaces (see e.g. [31]). Let $H_p(D)$, $p > 0$ be the one-dimensional complex quasi Banach space of analytic functions f on the unit

---

The research was supported by Shota Rustaveli National Science Foundation grant no. FR-21-2844.

G. Tephnadze (✉)
School of Science and Technology, The University of Georgia, Tbilisi, Georgia
e-mail: g.tephnadze@ug.edu.ge

R. Duduchava et al. (eds.), *Tbilisi Analysis and PDE Seminar*,
Research Perspectives Ghent Analysis and PDE Center 7,
https://doi.org/10.1007/978-3-031-62894-8_19

disc $D := (z : \ |z| < 1)$ for which

$$\|f\|_{H_p(D)} = \sup_{r<1} \frac{1}{2\pi} \left(\int_{[-\pi,\pi]} \left| f\left(re^{it}\right)\right|^p dt \right)^{1/p}.$$

The modulus of continuity in $H_p(D)$ $(0 < p < \infty)$ is defined by

$$\omega(\delta, f)_{H_p(D)} := \sup_{0\le h\le\delta} \left\| f(z) - f\left(ze^{ih}\right)\right\|_{H_p(D)}.$$

Now, we define real Hardy spaces. A real-valued distribution $f(t) \in D^{`}(T)$ belongs to the real Hardy space $H_p(T)$, where $0 < p < \infty$ and $T = (-\pi, \pi]$ if and only if there exists a function $F(z) \in H_p(D)$ with the properties $\operatorname{Im}(F(0)) = 0$ and

$$f(t) = \lim_{r\to 1} \operatorname{Re} F\left(re^{it}\right)$$

in the sense of distributions. The corresponding quasi-norm is defined by

$$\|f(z)\|_{H_p(T)} := \|F(z)\|_{H_p(D)}.$$

Hence, this class obviously becomes a real quasi-Banach space with quite the same properties as $H_p(D)$.

The definitions of n-dimensional Hardy spaces $H_p(D_n)$ $(p > 0)$ and real Hardy spaces $H_p(T_n)$ $(p > 0)$, where $D_n := D \times \cdots \times D$, $T_n := [-\pi, \pi)^n$ and theorems of atomic decompositions of these spaces can be found in Fefferman and Stein [4] (see also [31], Later [10] and Wilson [35]). Recall that $H_p(I_n) \sim L_p(I_n)$ for $p > 1$, where $I_n = D_n = T_n$. The rectangular partial sums with respect to the trigonometric system of function $f \in H_p(I_n)$ $p > 0$ is defined as follows

$$S_{N_1,\cdots,N_n}(f; x_1, \cdots, x_d) := \sum_{k_1=0}^{N_1} \cdots \sum_{k_n=0}^{N_n} c_{k_1,\cdots,k_n}(f)\, e^{i(k_1x_1+\cdots+k_nx_n)},$$

where $c_{k_1,\cdots,k_n}(f)$ are the Fourier coefficients of $f \in H_p(I_n)$, $(I_n = D_n,\ I_n = G_n)$ and modulus of continuity in $H_p(D_n)$ by $\omega(\delta, f)_{H_p(D_n)}$.

Next, we define the martingale Hardy spaces for $0 < p \le 1$ (see e.g. [15]). The dyadic (Walsh) group G is the complete direct product of the discrete cyclic group of order 2, that is $Z_2 := \{0, 1\}$, where the group operation is the modulo 2 addition, so G consists sequences $x = (x_i, i \in \mathbb{N})$, where $x_i \in Z_2$, $(i \in \mathbb{N})$. The group operation on G is the coordinate-wise addition: $x + y := \left(|x_n - y_n|, n \in \mathbb{N}\right)$.

It is well-known that the group G is a locally compact abelian group and the topology of G is the product topology, which has a countable base by the family

$$I_n(x) := \{y \in G : y_i = x_i, \ for \ 0 \leq i < n\}$$

and the measure μ is a Haar measure on G.

For every natural number $n = \sum_{i=0}^{\infty} n_i 2^i, \ n_i \in \{0, 1\}, \ i \in \mathbb{N}$, the $n-$th Walsh function is defined by

$$w_n(x) := \prod_{j=0}^{\infty} (-1)^{n_j x_j} = (-1)^{\sum_{j=0}^{\infty} n_j x_j} \quad (n \in \mathbb{N}).$$

The system $w := (w_n : n \in \mathbb{N})$ is called a Walsh system and it is orthonormal and complete in $L_2(G)$. Moreover, the set of Walsh functions coincides with the set of characters of the group G.

Let $G_n := G \times \ldots \times G$. The σ-algebra generated by the n-dimensional cubes $I_k^n(\overrightarrow{x}) = I_k(x_1) \times \ldots \times I_k(x_n)$, will be denoted by $F_k^n \ (k \in \mathbb{N})$.

Denote by $f = (f_n : n \in \mathbb{N})$ a one-parameter martingale with respect to σ-algebras $F_k^n \ (k \in \mathbb{N})$. For $0 < p < \infty$, the Hardy martingale space $H_p(G_n)$ consists of all martingales for which

$$\|f\|_{H_p(G_n)} := \|f^*\|_{L_p(G_n)} < \infty,$$

where

$$f^* := \sup_{n \in \mathbb{N}} |f_n|.$$

The rectangular partial sums of n-dimensional Walsh-Fourier series are defined as follows:

$$S_{(N_1,\ldots,N_n)}(f_1; \overrightarrow{x}) := \sum_{j_1=0}^{N_1-1} \cdots \sum_{j_n=0}^{N_n-1} \widehat{f}(j_1, \ldots, j_n) \prod_{i=1}^{n} w_{j_i}\left(x^i\right),$$

where $\widehat{f}(j_1, \ldots, j_n)$ are the Fourier coefficients of $f \in H_p(G_n)$.

The modulus of continuity in $H_p(G_n)$ $(0 < p < \infty)$ is defined by

$$\omega\left(1/2^k, f\right)_{H_p(G_n)} := \|f - S_{2^k,\ldots,2^k} f\|_{H_p(G_n)}.$$

Recall that $H_p(G_n) \sim L_p(G_n)$ for $p > 1$ (see [5] and [12]). Atomic decompositions of $H_p(G_n)$ $(0 < p \leq 1)$ spaces were proved by Weisz [32], [34] It is well-known that neither the Walsh nor the trigonometric system is a basis in L_1

(see e.g. [1], [8], [16], [36]). Moreover, (see e.g. [31], [32], [34], [27]) there exists a martingale in the Hardy space H_p, such that

$$\sup_n \|S_n f\|_{L_p} = \infty, \quad (0 < p \le 1).$$

The classical investigations by G. H. Hardy, F and M. Reisz, J. E. Littlewood and others of $H_p(D_n)$ (and $(H_p(T_n))$ employed complex methods and showed that, for several problems in Fourier Analysis, the use of the H_p scale $(0 < p \le 1)$ of spaces is preferable over that of L_p scale $(0 < p \le 1)$. During four decades a powerful theory of H_p $(0 < p \le 1)$ spaces, especially in the n-dimensional case, has been developed by means of real methods and various new applications to Fourier analysis and singular operators have been given e.g. by R. R. Coifman and G. Weiss (see [3]), C. Fefferman and E. M. Stein (see [4]), M. H. Taibleson and G. Weiss (see [24]) and the references given therein. However, concerning approximation properties (for instance, inequalities of Jackson type) only few results are known. We refer to the papers by Oswald [13], Kryakin and Trebels [9], Storoienko [19–23], where special integral representations and complex techniques have been used to obtain estimates of the following type: For example, for the one-dimensional trigonometric Fourier series Storozenko [20] proved the following inequality

$$\|S_k f - f\|_{H_p(D)} \le c_p \omega\Big(\frac{\pi}{n}, f\Big)_{H_p(D)} n^{1/p-1} (\ln n)^{[p]}.$$

It follows that if

$$\omega(\tfrac{\pi}{n}, f)_{H_p(D)} = O\left(1/\left(n^{1/p-1} (\ln n)^{[p]}\right)\right), \ as\ k \to \infty,$$

then

$$\|S_k f - f\|_{H_p(D)} \to 0, \ as\ k \to \infty.$$

For the one-dimensional Walsh-Fourier series in [26] it was proved that if

$$\omega\big(1/2^k, f\big)_{H_p(G)} = o\Big(1/(2^{k(1/p-1)} k^{[p]})\Big), \ as\ k \to \infty,$$

then the partial sums $S_n f$ of the one-dimensional Walsh system of $f \in H_p(G)$ converge to f in $H_p(G)$ norm, when $0 < p \le 1$, where $[p]$ denotes integer part of p. Moreover, there exists and martingale $f \in H_p(G)$ $0 < p \le 1$, for which

$$\omega\big(1/2^k, f\big)_{H_p(G)} = O\left(1/\left(2^{k(1/p-1)} k^{[p]}\right)\right), \ as\ k \to \infty$$

and

$$\overline{\lim_{k\to\infty}} \|S_k f - f\|_{H_p(G)} > c > 0.$$

The analogical results for the two-dimensional Walsh-Fourier series were proved in [14]. In particular, it was proved that if $0 < p \le 1,\ f \in H_p(I_n),\ 2^{-\alpha} \le k_1/k_2 \le 2^{\alpha}$ for some $\alpha > 0$ and

$$\omega\left(1/2^k, f\right)_{H_p(G_2)} = o\left(1/\left(2^{2k(1/p-1)}\log^{2[p]}k\right)\right), \quad \text{as} \quad k \to \infty.$$

Then

$$\left\| S_{k_1,k_2} f - f \right\|_{H_p(G_2)} \to 0, \quad \text{as} \quad k_1, k_2 \to \infty.$$

(b) (Sharpness) Let $0 < p \le 1,\ 2^{-\alpha} \le k_1/k_2 \le 2^{\alpha}$, for some $\alpha > 0$. Then there exists a martingale $f \in H_p(G_2)$, for which

$$\omega\left(1/2^k, f\right)_{H_p(G_2)} = O\left(1/\left(2^{2k(1/p-1)}(\log k)^{2[p]}\right)\right), \quad \text{as} \quad k \to \infty,$$

and

$$\overline{\lim_{k_1,k_2\to\infty}} \left\| S_{k_1,k_2} f - f \right\|_{H_p(G_2)} > c > 0.$$

Smith [18] (see also [2]) and Simon [17] the following strong summability result for one-parameter trigonometric-Fourier and Walsh-Fourier series, respectively,

$$\frac{1}{\log n} \sum_{k=1}^{n} \frac{\|S_k f\|_{H_1(I)}}{k} \le c\|f\|_{H_1(I)},\ f \in H_1(I)\ (n = 2, 3),$$

where $I = G$ or $I = D$. Moreover, Simon [17] and Weisz [32] proved that there exists an absolute constant depending only on p, such that

$$\sum_{k=1}^{\infty} \frac{\|S_k f\|^p{}_{H_p(I)}}{k^{2-p}} \le c_p \|f\|^p{}_{H_p(I)},\ f \in H_p(I) \quad (0 < p \le 1), \tag{1.1}$$

where $I = G$ or $I = T$. For the two-dimensional Walsh and trigonometric systems, Weisz [33] proved that if $0 < p \le 1$, $\alpha \ge 0$ and $f \in H_p(I_2)$, then there exists an absolute constant c_p depending only on p such that

$$\sup_{n,m\ge 2} \left(\frac{1}{\log n \log m}\right)^{[p]} \sum_{2^{-\alpha}\le k/l\le 2^{\alpha},(k,l)\le(n,m)} \frac{\left\|S_{k,l} f\right\|^p_{H_p(I_2)}}{(kl)^{2-p}} \le c_p \|f\|^p_{H_p(I_2)}, \tag{1.2}$$

where $I_2 = G_2$ or $I_2 = T_2$. Moreover, in [25, 30] it was proved that the rate of sequences $\left\{1/\left(k^{2-p}\right)\right\}_{k=1}^{\infty}$ and $\left\{1/\left((kl)^{2-p}\right)\right\}_{k=1,l=1}^{\infty}$, respectively, in (1.1) and (1.2) cannot be improved.

For the two-dimensional Walsh and trigonometric systems, Goginava and Gogoladze in [7] and [6] proved that for every $f \in H_1(I_2)$, there exists an absolute constant c, such that

$$\sum_{n=2}^{\infty}\frac{\left\|S_{n,n}f\right\|_{L_1(I_2)}}{n\log^2 n}\leq c\|f\|_{H_1(I_2)},$$

where $I_2 = G_2$ or $I_2 = T_2$.

In [11, 28] we generalized this result to the two-dimensional Walsh-Fourier series case when $0 < p \leq 1$ and proved that for every $f \in H_p(G_2)$, there exists an absolute constant c_p, depending only on p, such that

$$\sum_{n=2}^{\infty}\frac{\left\|S_{n,n}f\right\|^p_{L_p(G_2)}}{n^{3-2p}\log^{2[p]}n}\leq c_p\|f\|^p_{H_p(G_2)}.$$

Moreover, it was proved in [11, 29] that the rate of sequence $\left\{1/\left(k^{3-2p}\log^{2[p]}k\right)\right\}_{k=2}^{\infty}$ cannot be improved.

2 Main Results

Our first main result reads:

Theorem 2.1

(a) Let $0 < p \leq 1$, $f \in H_p(I_2)$, $2^{-\alpha} \leq k_i/k_{i+1} \leq 2^{\alpha}$ for some $\alpha > 0$ and

$$\omega(\pi/k, f)_{H_p(I_2)} = o\left(1/\left(k^{2(1/p-1)}(\log k)^{2[p]}\right)\right),\ as\ k\to\infty,$$

where $I_2 = D_2$, $I_2 = T_2$. Then

$$\left\|S_{k_1,k_2}f - f\right\|_{H_p(I_n)}\to 0,\ as\ k_1, k_2\to\infty.$$

(b) (Sharpness) Let $0 < p \leq 1$, $2^{-\alpha} \leq k_1/k_2 \leq 2^{\alpha}$, for some $\alpha > 0$. Then there exists a martingale $f \in H_p(I_2)$, for which

$$\omega(\pi/k, f)_{H_p(D_2)} = O\left(1/\left(k^{2(1/p-1)}(\log k)^{2[p]}\right)\right),\quad as\quad k\to\infty,$$

and

$$\overline{\lim_{k_1,k_2\to\infty}} \left\| S_{k_1,k_2} f - f \right\|_{H_p(I_2)} > c > 0,$$

where $I_2 = D_2$, $I_2 = T_2$.

The second main result is the following:

Theorem 2.2

(a) *Let* $f \in H_p(I_2)$, *where* $I_2 = D_2$, $I_2 = T_2$. *Then there exists an absolute constant* c_p *depending only on* p, *such that*

$$\sum_{n=2}^{\infty} \frac{\left\| S_{n,n} f \right\|^p_{L_p(I_2)}}{n^{3-2p} \log^{2[p]} n} \leq c_p \left\| f \right\|^p_{H_p(I_2)}.$$

Moreover, the rate of sequence $\left\{1/\left(k^{3-2p}\log^{2[p]}k\right)\right\}_{k=2}^{\infty}$ *cannot be improved.*

(b) *Let* $0 < p \leq 1$. *For any non-decreasing function* $\varphi(k) : \mathbb{N} \to \mathbb{R}$, *satisfying the condition*

$$\overline{\lim_{n\to\infty}} \left(k^{3-2p} \log^{2[p]} k\right) / \varphi(k) = \infty,$$

then there exists a martingale $f \in H_p(I_2)$, *where* $I_2 = D_2$, $I_2 = T_2$, *such that*

$$\sum_{k=2}^{\infty} \left\| S_{k,k} f \right\|^p_{H_p(I_2)} / \varphi(k) = \infty.$$

3 Open Problems and Final Remarks

First we present some natural open problems related to Theorems 2.1 and 2.2.

Problem 3.1 Does there exist necessary and sufficient conditions for the n-dimensional modulus of continuity in the classical and martingale Hardy spaces of $f \in H_p(I_n)$ $(0 < p \leq 1)$, $(I_n = G_n, I_n = D_n, I_n = T_n)$ for which

$$\left\| S_{k_1,k_2,\ldots,k_n} f - f \right\|_{H_p(I_n)} \to 0, \quad \text{as} \quad k_1, k_2, \ldots, k_n \to \infty,$$

in the case when the indices are restricted by the conditions $2^{-\alpha} \leq k_i/k_{i+1} \leq 2^{\alpha}$, $i = 1, \ldots, n-1$.

Problem 3.2 Is it possible to prove some corresponding strong summability theorems for the n-dimensional partial sums with respect to the Walsh and trigonometric systems in respective martingale and classical Hardy spaces H_p, for $0 < p \leq 1$, in the case when indices are restricted by the conditions $2^{-\alpha} \leq k_i/k_{i+1} \leq 2^{\alpha}$, $i = 1, \ldots, n-1$.

Problem 3.3 Is it possible to prove some corresponding strong summability theorems for the n-dimensional diagonal partial sums with respect to the Walsh and trigonometric systems in respective martingale and classical Hardy spaces H_p, for $0 < p \leq 1$.

Problem 3.4 Is it possible to prove some corresponding strong summability theorems for the n-dimensional partial sums without any restrictions on indices with respect to the Walsh and trigonometric systems in respective martingale and classical Hardy spaces H_p, for $0 < p \leq 1$.

Acknowledgments I would like to thank professor Lars-Erik Persson for his very good suggestions, which have improved the paper. The research was supported by The University of Georgia grant "Mathematical Analysis and PDE on Lie Groups."

References

1. G.N. Agaev, N.Ya. Vilenkin, G.M. Dzhafarly, A.I. Rubinshtein, *Multiplicative Systems of Functions and Harmonic Analysis on Zero-dimensional Groups* (Baku, Ehim, 1981)
2. E.S. Belinskii, Strong summability of Fourier series of the periodic functions from H_p $(0 < p < 1)$. Constr. Approx. **12**, 187–195 (1996)
3. R.R. Coifman, G. Weiss, Extensions of Hardy spaces and their use in analysis. Bull. Amer. Math. Soc. **83**, 569–645 (1977)
4. Ch. Fefferman, E.M. Stein, H^p spaces of several variables. Acta Math. **129**, 137–193 (1972)
5. A.M. Garsia, *Martingale Inequalities, Seminar Notes on Resent Progress*, Math. Lecture Notes Ser. (Benjamin, New York, 1973)
6. U. Goginava, L.D. Gogoladze, Strong convergence of cubic partial sums of two-dimensional Walsh-Fourier series, in *Constructive Theory of Functions, Sozopol 2010: In Memory of Borislav Bojanov* (Prof. Marin Drinov Academic Publishing House, Sofia, 2012), pp. 108–117
7. U. Goginava, L.D. Gogoladze, Strong summability theorems for $H_p(\Delta_d)$. Acta Math. Hungar. **138**(3), 259–266 (2013)
8. B. Golubov, A. Efimov, V. Skvortsov, *Walsh Series and Transformations* (Kluwer Acad. Publ., Dordrecht, 1991)
9. Y. Kryakin, W. Trebels, q-moduli of continuity in $H_p(D)$, $p > 0$ and an inequality of Hardy and Littlewood. J. Approx. Theory **115**, 238–259 (2002)
10. R.H. Later, A characterization of $H_p(R_n)$ in terms of atoms. Studia Math. **62**, 92–101 (1978)
11. N. Memić, I. Simon, G. Tephnadze, Strong convergence of two-dimensional Vilenkin-Fourier series. Math. Nachr. **289**(4), 485–500 (2016)
12. J. Neveu, *Discrete-Parameter Martingales* (North-Holland, 1971)
13. P. Oswald, On some approximation properties of real Hardy spaces $(0 < p \leq 1)$. J. Approx. Theory **40**, 45–65 (1984)
14. L.E. Persson, G. Tephnadze, P. Wall, On an approximation of two-dimensional Walsh-Fourier series in the martingale Hardy spaces. Ann. Funct. Anal. **9**(1), 137–150 (2018)
15. L.E. Persson, G. Tephnadze, F. Weisz, *Martingale Hardy Spaces and Summability of one-dimensional Vilenkin-Fourier Series* (Birkhäuser/Springer, 2022)

16. F. Schipp, W.R. Wade, P. Simon, J. Pál, *Walsh Series, An Introduction to Dyadic Harmonic Analysis* (Adam Hilger, Bristol, 1990)
17. P. Simon, Strong convergence of certain means with respect to the Walsh-Fourier series. Acta Math. Hungar. **49**(3–4), 425–431 (1987)
18. B. Smith, *A Strong Convergence Theorem for* $H_1(T)$. Lecture Notes in Math., vol. 995 (Springer, Berlin, 1994), pp. 169–173
19. E.A. Storozenko, On approximation of functions of the class H^p, $0 < p < 1$. Soobsc. Akad. Nauk Gruz. SSR **88**, 45–48 (1977) [Russian]
20. E.A. Storozenko, Approximation of functions of the class $H^p, 0 < p < 1$. Mat. Sb. **105**(149), 601–621 (1978) [Russian]
21. E.A. Storozenko, On the degree of approximation of functions of the class H^p, $0 < p < 1$. Dokl. Akad. Nauk Arm. SSR **66**, 145–149 [Russian] (1978)
22. E.A. Storozenko, Approximation of Functions and embedding Theorems in the Spaces HP and V, Dokt. Diss., Tbilisi State Univ., 1979 [Russian]
23. E.A. Storozenko, On theorems of Jackson type in H^p, $0 < p < 1$. Izv. Akad. Nauk SSSR Ser. Mat. **44**, 946–962 (1980) [Russianl]
24. M.H. Tableson, G. Weiss, The molecular characterization of certain Hardy spaces. Asterisque **77**, 68–149 (1980)
25. G. Tephnadze, A note on the Fourier coefficients and partial sums of Vilenkin-Fourier series. Acta Math. Acad. Paedagog. Nyházi. **28**(2), 167–176 (2012)
26. G. Tephnadze, On the partial sums of Vilenkin-Fourier series. J. Contemp. Math. Anal. **49**(1), 23–32 (2014)
27. G. Tephnadze, Martingale Hardy Spaces and Summability of the One-Dimensional Vilenkin-Fourier Series, PhD thesis, Department of Engineering Sciences and Mathematics, Luleå University of Technology, Oct. 2015 (ISSN 1402–1544)
28. G. Tephnadze, Strong convergence of two-dimensional Walsh-Fourier series. Ukr. Math. J. **65**(6), 914–926 (2013)
29. G. Tephnadze, A note on the strong convergence of two-dimensional Walsh-Fourier series. Proc. Razmadze Math. Inst. **162**, 93–97 (2013)
30. G. Tephnadze, A note on strong summability of two-dimensional Walsh-Fourier series. Georgian Math. J. **28**(3), 477–482 (2021)
31. A. Torchinsky, *Real-Variable Methods in Harmonic Analysis*, Dover Books on Mathematics (2008)
32. F. Weisz, *Martingale Hardy Spaces and their Applications in Fourier Analysis* (Springer, Berlin, 1994)
33. F. Weisz, Strong convergence theorems for two-parameter Walsh-Fourier and trigonometric-Fourier series. (English) Stud. Math. **117**(2), 173–194 (1996)
34. F. Weisz, *Summability of Multi-dimensional Fourier Series and Hardy Space*, Math. Appl. (Kluwer Academic, Dordrecht, 2002)
35. J.M. Wilson, A simple proof of the atomic decomposition for $H_p(R_n)$ $0 < p \leq 1$. Studia Math. **74**, 25–33 (1982)
36. A. Zygmund, *Trigonometric Series*, vol. 1 (Cambridge University Press, 1959)

Modulus of Continuity and Convergence of Fejér Means of Vilenkin-Fourier Series in the Variable Martingale Hardy Space $H_{p(\cdot)}$

George Tephnadze and Ferenc Weisz

Abstract The main aim of this paper is to investigate the weighted maximal operator $\sup_{k\in\mathbb{N}}\left(|\sigma_k f|/\log^2(k+1)\right)$ of Fejér means of Vilenkin-Fourier series and prove that the it is bounded from the variable martingale Hardy space $H_{p(\cdot)}$ to the Lebesgue space $L_{p(\cdot)}$, where $p(\cdot)\geq \underline{p}\geq 1/2$. By using this result we get sufficient condition for the modulus of continuity which is guarantee for the convergence of Fejér means in the variable martingale Hardy space $H_{p(\cdot)}$. We also show that this condition is sharp. As a consequence we obtain some new and well-know results.

Keywords Vilenkin system · Fejér means · Variable martingale Hardy space · Maximal operator

2010 Mathematics Subject Classification 42C10

1 Introduction

In the one-dimensional case the weak $(1, 1)$-type inequality for the maximal operator of Fejér means with respect to Vilenkin system

$$\sigma^* f := \sup_{n\in\mathbb{N}} |\sigma_n f|$$

can be found in Schipp [24] and Pál, Simon [19] (see also [2] and [15]). Fujii [11] and Simon [26] verified that σ^* is bounded from H_1 to L_1. Weisz [35] (see

G. Tephnadze
School of Science and Technology, The University of Georgia, Tbilisi, Georgia
e-mail: g.tephnadze@ug.edu.ge

F. Weisz (✉)
Department of Numerical Analysis, Eötvös University, Budapest, Hungary
e-mail: weisz@inf.elte.hu

R. Duduchava et al. (eds.), *Tbilisi Analysis and PDE Seminar*,
Research Perspectives Ghent Analysis and PDE Center 7,
https://doi.org/10.1007/978-3-031-62894-8_20

also [21] and [23]) generalized this result and proved boundedness of σ^* from the martingale space H_p to the Lebesgue space L_p for $p > 1/2$. Simon [27] gave a counterexample, which shows that boundedness does not hold for $0 < p < 1/2$. A counterexample for $p = 1/2$ was given by Goginava [13]. Moreover, [14] (see also [28]) he proved that there exists a martingale $F \in H_p$ $(0 < p \leq 1/2)$, such that $\sup_{n\in\mathbb{N}} \|\sigma_n F\|_p = +\infty$. Weisz [37] proved that the maximal operator σ^* of the Fejér means is bounded from the Hardy space $H_{1/2}$ to the space $weak - L_{1/2}$.

In [29] was proved that the weighted maximal operator

$$\overset{\sim}{\sigma}^* f := \sup_{n\in\mathbb{N}} \frac{|\sigma_n f|}{\log^2(n+1)} \tag{1.1}$$

is bounded from the Hardy space $H_{1/2}(G)$ to the space $L_{1/2}(G)$. Moreover, it was also proved that the rate of denominator $\{\log^2(n+1)\}$ can not be improved.

In [30, 32] (see also [22, 31, 33]) it was proved that if $f \in H_{1/2}$ and

$$\omega_{H_{1/2}}\left(\frac{1}{M_N}, f\right) = o\left(\frac{1}{N^2}\right), \quad \text{as } N \to \infty, \tag{1.2}$$

then

$$\|\sigma_n f - f\|_{H_{1/2}} \to 0, \text{ as } n \to \infty.$$

Moreover, it was also proved that there exists a martingale $f \in H_{1/2}(G_m)$ for which

$$\omega_{H_{1/2}}\left(\frac{1}{M_N}, f\right) = O\left(\frac{1}{N^2}\right), \quad \text{as } N \to \infty \tag{1.3}$$

and

$$\|\sigma_n f - f\|_{1/2} \nrightarrow 0, \quad \text{as } n \to \infty.$$

Similar problems for general summability methods with respect to the one- and two-dimensional cases were studied by a lot of authors. We mention Baramidze, Blahota, Gát, Goginava, Nagy, Persson, Tephnadze, Toledo, Weisz [1, 3–8, 12, 14, 16–18, 20, 23].

Theory of Hardy and Lebesgue spaces with variable exponents is investigated very intensively in the literature nowadays. It needs essentially new ideas. Interesting problems in this theory were studied by a lot of authors. We mention for instance Cruz-Uribe and Fiorenza, Diening, Kempka and Vybíral, Nakai and Sawano, Jiao, Yan and Liu. The interest in the variable Lebesgue spaces has increased since the 1990s because of their use in a variety of applications, such as in fluid dynamics, image processing, partial differential equations, variational calculus and harmonic analysis.

In [36] (see also [34]) Weisz proved that if $p(\cdot) \in C^{\log}$ and $1/2 \leq \underline{p}$ and

$$\frac{1}{p_-} - \frac{1}{p_+} < 1, \tag{1.4}$$

then

$$\|Tf\|_{p(\cdot)} \leq C_{p(\cdot)} \|f\|_{H_{p(\cdot)}} \qquad (f \in H_{p(\cdot)}).$$

The main aim of this paper is to prove that the maximal operator (1.1) is bounded from is bounded from the variable martingale Hardy space $H_{p(\cdot)}$ to the Lebesgue space $L_{p(\cdot)}$, where $p(\cdot) \geq \underline{p} \geq 1/2$. By using this result we get sufficient condition for the modulus of continuity which is guarantee for the convergence of Fejér means in the variable martingale Hardy space $H_{p(\cdot)}$. We also show that this condition is sharp. As a consequence we obtain some new and well-know results.

2 Definitions and Notations

Let $\mathbb{N}_+$ denote the set of the positive integers, $\mathbb{N} := \mathbb{N}_+ \cup \{0\}$. Let us define by $\mathbb{Q}_m$ the set of rational numbers of the form pM_n^{-1}, where $0 \leq p \leq M_n - 1$ for some $p \in \mathbb{N}$ and $n \in \mathbb{N}$. Let $m := (m_{0,} m_1, \ldots)$ denote a sequence of the positive integers not less than 2. If we define the so-called generalized number system based on m as $M_0 := 1,\ M_{k+1} := m_k M_k\ \ (k \in \mathbb{N})$ then every $n \in \mathbb{N}$ can be uniquely expressed as $n = \sum_{k=0}^{\infty} n_j M_j$, where $n_j \in Z_{m_j}$, $(j \in \mathbb{N})$ and only a finite number of n_j's differ from zero. Let $|n| := \max\{j \in \mathbb{N},\ n_j \neq 0\}$.

Any $x \in [0, 1)$ can be written in the form

$$x = \sum_{k=0}^{\infty} \frac{x_k}{M_{k+1}}, \tag{2.1}$$

where each $x_k \in Z_{m_k}$. For each $x \in [0, 1) \setminus \mathbb{Q}_m$ there is only one expression of this form. When $x \in \mathbb{Q}_m$ there are two expressions of this form, one terminates in 0's and one does not terminate in 0's. By expansion (2.1) of an element $x \in \mathbb{Q}_m$ we shall mean the one which terminates in 0's.

The Rademacher and Vilenkin functions are given by generalized Rademacher functions

$$r_n(x) = \exp\left(2\pi \imath x_k / m_k\right) \qquad \text{and} \qquad \psi_n =: \prod_{k=0}^{\infty} r_k^{n_k} \qquad (n \in \mathbb{N}),$$

where $x \in [0, 1)$ has the expansion (2.1) and $n = \sum_{j=0}^{\infty} n_j M_j$.

The Vilenkin system is orthonormal and complete in $L_2(G)$ (see [25]).

We recall some basic notations on variable Lebesgue spaces and give some elementary and necessary facts about these spaces. The proofs are omitted because they can be found in the excellent books Cruz-Uribe and Fiorenza [9] and Diening et al. [10].

A measurable function $p(\cdot) : G_m \to (0, \infty)$ is called a variable exponent. For any variable exponent $p(\cdot)$ and any measurable set $A \subset G_m$, we will use the notation

$$p_-(A) := \operatorname*{ess\,inf}_{x\in A} p(x) \qquad \text{and} \qquad p_+(A) := \operatorname*{ess\,sup}_{x\in A} p(x).$$

If $A = G_m$, then the numbers $p_-(A)$ and $p_+(A)$ are denoted simply by p_- and p_+. Denote by $\mathbb{V}$ the collection of all variable exponents $p(\cdot)$ satisfying $0 < p_- \leq p_+ < \infty$. In what follows, we use the symbol $\underline{p}$, where $\underline{p} := \min\{p_-, 1\}$.

Definition 2.1 For $p(\cdot) \in \mathbb{V}$ and a measurable function f, the modular functional $\rho_{p(\cdot)}$ is defined by

$$\rho_{p(\cdot)}(f) := \int_0^1 |f(x)|^{p(x)} \, d\mu(x)$$

and the Luxemburg quasi-norm is given by setting

$$\|f\|_{p(\cdot)} := \inf\left\{\eta \in (0, \infty) : \rho_{p(\cdot)}\left(\frac{f}{\eta}\right) \leq 1\right\}.$$

The variable Lebesgue space $L^{p(\cdot)}$ is defined to be the set of all measurable functions f such that $\rho_{p(\cdot)}(f) < \infty$ and equipped with the quasi-norm $\| \cdot \|_{p(\cdot)}$.

It is easy to see that if $p(\cdot)$ is a constant, then we get back the L^p spaces. For $p(\cdot) \in \mathbb{V}$ and $f, g \in L^{p(\cdot)}$, we can easily prove that

$$\|\eta f\|_{p(\cdot)} = |\eta| \|f\|_{p(\cdot)} \qquad (\eta \in \mathbb{C}),$$

$$\left\| |f|^s \right\|_{p(\cdot)} = \|f\|^s_{sp(\cdot)} \qquad (s \in (0, \infty))$$

and

$$\|f + g\|^{\underline{p}}_{p(\cdot)} \leq \|f\|^{\underline{p}}_{p(\cdot)} + \|g\|^{\underline{p}}_{p(\cdot)}.$$

The last inequality is a generalization of Minkowski's inequality. Details can be found in the monographs Cruz-Uribe and Fiorenza [9] and Diening et al. [10]. The function $p'(\cdot)$ denotes the conjugate exponent function of $p(\cdot)$, i.e.,

$$\frac{1}{p(x)} + \frac{1}{p'(x)} = 1 \qquad (x \in G_m).$$

The well known Hölder's inequality can be generalized for variable Lebesgue spaces (see Cruz-Uribe and Fiorenza [9, p.27] or Diening et al. [10, p.74]).

Theorem 2.2 *Let $p(\cdot) \in \mathbb{V}$ with $p_- \geq 1$. Then, for all $f \in L^{p(\cdot)}$ and $g \in L^{p'(\cdot)}$,*

$$\int_{G_m} |fg| \, d\mu \leq C_{p(\cdot)} \|f\|_{p(\cdot)} \|g\|_{p'(\cdot)} .$$

Definition 2.3 We say that a function $p(\cdot) \in \mathbb{V}$ satisfies the log-Hölder continuity condition, if there exists a positive constant $C_{\log}(p)$ such that, for any $x, y \in [0, 1)$,

$$|p(x) - p(y)| \leq \frac{C_{\log}(p)}{\log(e + 1/|x - y|)}. \tag{2.2}$$

The set of all these functions is denoted by $C^{\log}$.

The following two lemmas were proved in [9] and [10]:

Lemma 2.4 *We have that $p(\cdot) \in C^{\log}$ if and only if there exists $K > 1$ such that for all intervals $I \subset [0, 1)$,*

$$\lambda(I)^{p_-(I)-p_+(I)} \leq K. \tag{2.3}$$

Lemma 2.5 *If $p(\cdot) \in C^{\log}$, then, for any interval $I \subset [0, 1)$ and any $x \in I$,*

$$\lambda(I)^{1/p_-(I)} \sim \lambda(I)^{1/p(x)} \sim \lambda(I)^{1/p_+(I)} \sim \|\chi_I\|_{p(\cdot)},$$

where $\sim$ denotes the equivalence of the numbers.

Remark 2.6 There exist a lot of functions $p(\cdot)$ satisfying (2.2). For concrete examples we mention the function $a + cx$ with parameters a and c such that the function is positive ($x \in [0, 1)$). All positive Lipschitz functions with order $0 < \alpha \leq 1$ also satisfy (2.2) and (2.3).

It is clear that (2.3) implies that there exists $0 < \beta < 1$ such that

$$\lambda(I)^{1/p_+(I)-1/p_-(I)} \leq \frac{1}{\beta}. \tag{2.4}$$

Actually, $\beta = K^{-1/(p_+(I)p_-(I))}$ and (2.3) is equivalent to (2.4) because $0 < p_- \leq p_+ < \infty$. Note that K and β are depending on $p(\cdot)$.

If $f \in L_1[0,1)$, we can establish Fourier coefficients, partial sums of Fourier series, Fejér means in the usual manner:

$$\widehat{f}(n) := \int_G f\overline{\psi_n}d\mu, \quad (n \in \mathbb{N}),$$

$$S_n f := \sum_{k=0}^{n-1} \widehat{f}(k)\,\psi_k, \quad (n \in \mathbb{N}_+, S_0 f := 0),$$

$$\sigma_n f := \frac{1}{n}\sum_{k=1}^{n} S_k f.$$

we generalize Doob's inequality to the variable Lebesgue space $L^{p(\cdot)}$. A martingale $f = (f_n)_{n\geq 0}$ is called an $L_{p(\cdot)}$-martingale if $f_n \in L^{p(\cdot)}$ for all $n \in N$. In this case, we set

$$\|f\|_{L_{p(\cdot)}} = \sup_{n\geq 0} \|f_n\|_{p(\cdot)}.$$

If $\|f\|_{L_{p(\cdot)}} < \infty$, then f is said to be a $L_{p(\cdot)}$-bounded martingale and it is denoted by $f \in L_{p(\cdot)}$. For a martingale, we define Doob's maximal functions M_m, $m \in \mathbb{N}$, and M of f by $L_{p(\cdot)}$

$$M_m(f) = \sup_{0\leq n\leq m} |f_n|, \qquad M(f) = \sup_{n\geq 0} |f_n|.$$

The maximal function $M(f)$ is also denoted by f^*. For an integrable function f, we define the maximal operator in the same way as for the martingale $(E_n f)$.

Doob's inequality, for a constant p, $\|M(f)\|_p \leq C_p \|f\|_p$ $(1 < p < \infty)$ can be generalized to variable Lebesgue spaces as follows.

Theorem 2.7 *Let $p(\cdot) \in C^{\log}$ and $f = (f_n) \in L_{p(\cdot)}$. If $1 < p_- \leq p_+ < \infty$, then Doob's inequality*

$$\|M(f)\|_{p(\cdot)} \leq C_{p(\cdot)} \|f\|_{L_{p(\cdot)}}. \tag{2.5}$$

If $1 \leq p_- \leq p_+ < \infty$, then

$$\sup_{\lambda>0} \left\|\lambda\chi_{\{M(f)>\lambda\}}\right\|_{p(\cdot)} \leq C_{p(\cdot)} \|f\|_{L_{p(\cdot)}}. \tag{2.6}$$

Let us first recall some definitions. For a martingale $f = (f_n)_{n\geq 0}$,

$$d_n f = f_n - f_{n-1} \qquad (n \geq 0)$$

denote the martingale differences, where $f_{-1} = 0$. We may suppose also that $f_0 = 0$. For a martingale, we define the square functions S_m and S and the conditional square functions s_m and s of f, respectively, as follows:

$$S_m(f) := \left(\sum_{n=0}^{m} |d_n f|^2\right)^{1/2}, \quad S(f) := \left(\sum_{n=0}^{\infty} |d_n f|^2\right)^{1/2}$$

and

$$s_m(f) := \left(\sum_{n=0}^{m} E_{n-1}|d_n f|^2\right)^{1/2}, \quad s(f) := \left(\sum_{n=0}^{\infty} E_{n-1}|d_n f|^2\right)^{1/2},$$

where $E_{-1} := E_0$. For an integrable function f, we define the operators in the same way as for the martingale $(E_n f)$. $S(f)$ $s(f)$ $H^M_{p(\cdot)}$ $H^S_{p(\cdot)}$ $H^s_{p(\cdot)}$

Definition 2.8 The variable martingale Hardy spaces $H^M_{p(\cdot)}$, $H^S_{p(\cdot)}$ and $H^s_{p(\cdot)}$ associated with the variable Lebesgue spaces $L^{p(\cdot)}$ are defined by

$$H^M_{p(\cdot)} := \{f = (f_n)_{n\geq 0} : \|f\|_{H^M_{p(\cdot)}} = \|M(f)\|_{L^{p(\cdot)}} < \infty\},$$

$$H^S_{p(\cdot)} := \{f = (f_n)_{n\geq 0} : \|f\|_{H^S_{p(\cdot)}} = \|S(f)\|_{L^{p(\cdot)}} < \infty\}$$

$$H^s_{p(\cdot)} := \{f = (f_n)_{n\geq 0} : \|f\|_{H^s_{p(\cdot)}} = \|s(f)\|_{L^{p(\cdot)}} < \infty\}.$$

If $p(\cdot) = p$ is a constant, then the above definitions of variable Hardy spaces coincide with the classical definitions in [23] and [34].

Corollary 2.9 *If $p(\cdot) \in C^{\log}$ and $1 < p_- \leq p_+ < \infty$, then $H^M_{p(\cdot)}$ is equivalent to $L_{p(\cdot)}$ with the inequalities*

$$\|f\|_{L_{p(\cdot)}} \leq \|f\|_{H^M_{p(\cdot)}} \leq C_{p(\cdot)} \|f\|_{L_{p(\cdot)}}.$$

Next, we will see that the Lebesgue space $L^{p(\cdot)}$ is also equivalent to these spaces.

Corollary 2.10 *Let $p(\cdot) \in C^{\log}$. If $1 < p_- \leq p_+ < \infty$, then $H^M_{p(\cdot)}$ is equivalent to $L^{p(\cdot)}$ with the inequality $\|f\|_{p(\cdot)} \leq \|f\|_{L_{p(\cdot)}} \leq \|f\|_{H^M_{p(\cdot)}} \leq C_{p(\cdot)} \|f\|_{p(\cdot)}$.*

If $1 \leq p_- \leq p_+ < \infty$, then $H^M_{p(\cdot)} \subset L^{p(\cdot)}$ with the inequality

$$\|f\|_{p(\cdot)} \leq \|f\|_{H^M_{p(\cdot)}}.$$

3 Formulation of Main Results

Theorem 3.1 *Let $p(\cdot) \in C^{\log}$ and $t < 1/2 \leq \underline{p}$. If condition* (1.4) *is fulfilled, then the maximal operator*

$$\overset{\sim}{\sigma}^{*} f := \sup_{n \in \mathbb{N}} \frac{|\sigma_n f|}{\log^2 (n+1)}. \tag{3.1}$$

is bounded from the variable Hardy space $H_{p(\cdot)}$ to the space $L_{p(\cdot)}$.

In the concrete case when $p_- \leq p(\cdot) \leq p_+ = 1/2$, we get that the following:

Corollary 3.2 *Let $f \in H_{1/2}(G_m)$. Then the maximal operator, defined by* (3.1), *is bounded from the Hardy space $H_{1/2}(G)$ to the space $L_{1/2}(G)$.*

Remark 3.3 In [32] it was also proved that for any nondecreasing sequence $\{\varphi_n\}$ satisfying the condition

$$\overline{\lim_{n \to \infty}} \frac{\log^2 (n+1)}{\varphi_n} = +\infty.$$

there exists a martingale $f \in H_{1/2}$, such that the maximal operator $\sup_{n \in \mathbb{N}} \frac{|\sigma_n f|}{\varphi_n}$ is not bounded from the Hardy space $H_{1/2}(G)$ to the Lebesgue space $L_{1/2}(G)$, which follows that the rate of the sequence $\{\log^2 (n+1)\}$ can not be improved in Corollary 3.2 and Theorem 3.1.

Theorem 3.4 *Let $p(\cdot) \in C^{\log}$ and $t < 1/2 \leq \underline{p}$ and $f \in H_{p(\cdot)}$. If condition* (1.4) *is fulfilled and*

$$\omega_{H_{p(\cdot)}} \left(\frac{1}{M_N}, f \right) = o \left(\frac{1}{N^2} \right), \quad as \ N \to \infty, \tag{3.2}$$

then

$$\|\sigma_n f - f\|_{H_{p(\cdot)}} \to 0, \quad as \ n \to \infty.$$

In the concrete case when $p_- \leq p(\cdot) \leq p_+ = 1/2$, we get that the following:

Corollary 3.5 *Let $f \in H_{1/2}$, and*

$$\omega_{H_{1/2}} \left(\frac{1}{M_N}, f \right) = o \left(\frac{1}{N^2} \right), \quad as \ N \to \infty. \tag{3.3}$$

Then

$$\|\sigma_n f - f\|_{H_{1/2}} \to 0, \quad as \ n \to \infty.$$

Remark 3.6 In [31] it was proved that there exists a martingale $f \in H_{1/2}(G_m)$ for which

$$\omega_{H_{1/2}}\left(\frac{1}{M_N}, f\right) = O\left(\frac{1}{N^2}\right), \quad \text{as } N \to \infty \tag{3.4}$$

and

$$\|\sigma_n f - f\|_{H_{1/2}} \nrightarrow 0, \quad \text{as } n \to \infty,$$

which follows that the rate of the sequence $\{\log^2(n+1)\}$ can not be improved in Corollary 3.5 and Theorem 3.4.

Acknowledgments The research was supported by The University of Georgia grant "Mathematical Analysis and PDE on Lie Groups".

References

1. D. Baramidze, I. Blahota, G. Tephnadze, R. Toledo, Martingale Hardy spaces and some new weighted maximal operator of Fejér means of Walsh-Fourier series. J. Geom. Anal. (2024), 34, paper no. 3, 17 pp.
2. D. Baramidze, N. Nadirashvili, L.-E. Persson, G. Tephnadze, Some weak-type inequalities and almost Everywhere convergence of Vilenkin-Nörlund means. J. Inequal. Appl. (2023), paper no. 66, 17 pp.
3. D. Baramidze, L. Baramidze, L.E. Persson, G. Tephnadze, Some now restricted maximal operators of Fejér means of Walsh-Fourier series. Banach J. Math. Anal. **17**(4), 75 (2023)
4. I. Blahota, G. Tephnadze, On the (C, α)-means with respect to the Walsh system. Anal. Math. **40**, 161–174 (2014)
5. I. Blahota, G. Tephnadze, Strong convergence theorem for Vilenkin-Fejér means. Publ. Math. Debrecen **85**(1–2), 181–196 (2014)
6. I. Blahota, L.E. Persson, G. Tephnadze, On the Nörlund means of Vilenkin-Fourier series. Czech. Math J. **65**(4), 983–1002 (2015)
7. I. Blahota, G. Tephnadze, R. Toledo, Strong convergence theorem of (C, α)-means with respect to the Walsh system. Tohoku Math. J. **67**(4), 573–584 (2015)
8. I. Blahota, K. Nagy, G. Tephnadze, Approximation by Marcinkiewicz Θ-means of double Walsh-Fourier series. Math. Inequal. Appl. **22**(3), 837–853 (2019)
9. D. Cruz-Uribe, A. Fiorenza, *Variable Lebesgue Spaces: Foundations and Harmonic Analysis*, Applied and Numerical Harmonic Analysis (Birkhäuser, Springer, Heidelberg, 2013)
10. L. Diening, P. Harjulehto, P. Hästö, M. Ruzicka, *Lebesgue and Sobolev spaces with Variable Exponents, 2011*, Lecture Notes in Mathematics (Springer, Heidelberg, 2017)
11. N.J. Fujii, A maximal inequality for H_1 functions on the generalized Walsh-Paley group. Proc. Amer. Math. Soc. **77**, 111–116 (1979)
12. G. Gát, Inverstigations of certain operators with respect to the Vilenkin system. Acta Math. Hung. **61**, 131–149 (1993)
13. U. Goginava, Maximal operators of Fejér means of double Walsh-Fourier series. Acta Math. Hungar. **115**, 333–340 (2007)

14. U. Goginava, The martingale Hardy type inequality for Marcinkiewicz-Fejér means of two-dimensional conjugate Walsh-Fourier series. Acta Math. Sin. **27**, 1949–1958 (2011)
15. B. Golubov, A. Efimov, V. Skvortsov, *Walsh Series and Transformations* (Kluwer Acad. Publ, Dordrecht, 1991)
16. K. Nagy, G. Tephnadze, Walsh-Marcinkiewicz means and Hardy spaces. Cent. Eur. J. Math. **12**(8), 1214–1228 (2014)
17. K. Nagy, G. Tephnadze, Approximation by Walsh-Marcinkiewicz means on the Hardy space. Kyoto J. Math. **54**(3), 641–652 (2014)
18. K. Nagy, G. Tephnadze, The Walsh-Kaczmarz-Marcinkiewicz means and Hardy spaces. Acta Math. Hung. **149**(2), 346–374 (2016)
19. J. Pál, P. Simon, On a generalization of the concept of derivate. Acta Math. Hung. **29**, 155–164 (1977)
20. L.E. Persson, G. Tephnadze, A sharp boundedness result concerning some maximal operators of Vilenkin-Fejér means. Mediterr. J. Math. **13**(4), 1841–1853 (2016)
21. L.E. Persson, G. Tephnadze, P. Wall, On the maximal operators of Vilenkin-Nörlund means. J. Fourier Anal. Appl. **21**(1), 76–94 (2015)
22. L.E. Persson, G. Tephnadze, G. Tutberidze, On the boundedness of subsequences of Vilenkin-Fejér means on the martingale Hardy spaces. Oper. Matrices **14**(1), 283–294 (2020)
23. L.E. Persson, G. Tephnadze, F. Weisz, *Martingale Hardy Spaces and Summability of Vilenkin-Fourier Series*, book manuscript (Birkhäuser/Springer, 2022)
24. F. Schipp, Certain rearrangements of series in the Walsh series. Mat. Zametki **18**, 193–201 (1975)
25. F. Schipp, W. Wade, P. Simon, J. Pál, *Walsh Series, An Introduction to Duadic Harmonic Analysis* (Akademiai Kiadó, Budapest-AdamHilger, 1990)
26. P. Simon, Investigations with respect to the Vilenkin system. Ann. Univ. Sci. Budapest. Eötvös Sect. Math. **28**, 87–101 (1985)
27. P. Simon, Cesáro summability with respect to two-parameter Walsh systems. Monatsh. Math. **131**(4), 321–334 (2000)
28. G. Tephnadze, Fejér means of Vilenkin-Fourier series. Stud. Sci. Math. Hung. **49**, 79–90 (2012)
29. G. Tephnadze, On the maximal operator of Vilenkin-Fejér means. Turk. J. Math. **37**, 308–318 (2013)
30. G. Tephnadze, A note on the norm convergence by Vilenkin-Fejér means. Georgian Math. J. **21**(4), 511–517 (2014)
31. G. Tephnadze, On the partial sums of Vilenkin-Fourier series. J. Contemp Math. Anal. **49**(1), 23–32 (2014)
32. G. Tephnadze, On the convergence of Fejér means of Walsh-Fourier series in the space H_p. J. Contemp. Math. Anal. **51**(2), 90–102 (2016)
33. G. Tephnadze, The one-dimensional martingale Hardy spaces and partial sums and Fejér means with respect to Walsh system. Mem. Differential Equations Math. Phys. **88**, 109–158 (2023)
34. F. Weisz, *Martingale Hardy spaces and their applications in Fourier Analysis* (Springer, Berlin, 1994)
35. F. Weisz, Cesáro summability of one- and two-dimensional Walsh-Fourier series. Anal. Math. **22**, 229–242 (1996)
36. F. Weisz, *Summability of Multi-dimensional Fourier Series and Hardy Space* (Kluwer Academic, Dordrecht, 2002)
37. F. Weisz, Weak type inequalities for the Walsh and bounded Ciesielski systems. Anal. Math. **30**(2), 147–160 (2004)

On Unconditional Convergence of Fourier Series with Respect to General Orthonormal Systems

V. Tsagareishvili, G. Tutberidze, and G. Cagareishvili

Abstract S. Banach, in particular, proved that for any function, even $f(x) = 1$, where $x \in [0, 1]$, the convergence of its Fourier series with respect to the general orthonormal systems (ONS) is not guaranteed. We find conditions for the functions φ_n of an ONS (φ_n), under which the Fourier series of functions $f \in Lip_1$ are unconditionally convergent almost everywhere. During our research, we mainly used the properties of the sequences of linear functionals on the Banach spaces to prove the main theorems we presented in this article. Our research has concluded that the aforementioned conditions do exist and are the best possible in a certain sense. We have also found that any ONS contains a subsystem such that the Fourier series of any function $f \in Lip_1$ is unconditionally convergent. Further, the precondition presented in Theorem 2, which demands that the Fourier series of the function $f = 1$ be convergent, is adequate, meaning that it does not make the actual conditions of Theorem 2 redundant. We have shown that the solution for these types of problems for the general ONS is trivial for the classical ONS (trigonometric, Haar, and Walsh systems).

Keywords Unconditional convergence · Orthonormal systems · Fourier coefficients · Sequence of linear functionals · Banach space

2020 Mathematics Subject Classification 42C10, 46B07

V. Tsagareishvili · G. Cagareishvili
Department of Mathematics, Faculty of Exact and Natural Sciences, Ivane Javakhishvili Tbilisi State University, Tbilisi, Georgia

G. Tutberidze (✉)
School of Science and Technology, The University of Georgia, Tbilisi, Georgia
e-mail: g.tutberidze@ug.edu.ge

R. Duduchava et al. (eds.), *Tbilisi Analysis and PDE Seminar*,
Research Perspectives Ghent Analysis and PDE Center 7,
https://doi.org/10.1007/978-3-031-62894-8_21

1 Introductions

In order not to disturb the discussion in this introduction and our main result we have collected all notations, definitions and other preliminaries in Sect. 2.

In this paper we find conditions for the functions φ_n of an ONS (φ_n), under which the Fourier series of functions $f \in Lip_1$ are unconditionally convergent almost everywhere (See Theorem 3.1). We also find that this result is, in a sense, sharp (See Theorem 3.2). Furthermore, we discover that any ONS contains a subsystem with respect to which the general Fourier series of any function is unconditionally convergent. Lastly, we clarify that the precondition imposed upon the functions of the ONS is accurate and valid, meaning that it does not make the actual condition presented in Theorem 3.1 redundant.

In the monographs [1, 12, 13] and papers [17, 18, 24, 26] problems of the theory of convergence of orthonormal series are studied. It is interesting that according to Menchov's and Banach's results the convergence of general orthonormal series and the convergence of general Fourier series for functions from some differential class are different problems. In the first case, the coefficients of the orthonormal series play the decisive role. In the second case, the fact that a function f $(f \neq 0)$ belongs to any differential class does not guarantee the convergence of its Fourier series with respect to general ONS. Therefore, for the Fourier series, with respect to an ONS, to be convergent, one should impose additional conditions on the functions φ_n of the ONS (φ_n). Our main problem is to find these conditions for the functions φ_n of an ONS (φ_n) for the Fourier series of every function of Lip_1 classes to be unconditionally convergent.

Similar problems are studied in the papers [3–10, 14–16, 19–23, 25, 27–38].

The main results of Theorems 3.1–3.6 and Corollary 3.3 are presented in Sect. 3. The new applications concerning the convergence of the general Fourier series can be found in Sect. 4. See Theorems 4.1 and 4.2.

2 Preliminaries

Let (φ_n) be ONS on $[0, 1]$. Suppose that $\varepsilon \in (0, 1)$, $p = p(\varepsilon) = 2 - \varepsilon$ and $q = q(\varepsilon) = \frac{2-\varepsilon}{1-\varepsilon}$.

Firstly, we must establish the following expression

$$M_n(a, \varepsilon) := \frac{1}{n} \sum_{i=1}^{n-1} \left| \int_0^{i/n} P_n(a, \varepsilon, x)dx \right|, \tag{2.1}$$

where $(a_n) \in l_{q(\varepsilon)}$ and

$$P_n(a, \varepsilon, x) := \sum_{k=1}^{n} a_k \varphi_k(x). \tag{2.2}$$

Let us suppose that $f \in L_2$, then the sequence of real numbers

$$C_n(f) = \int_0^1 f(x)\varphi_n(x)dx, \quad (n = 1, 2, \dots) \tag{2.3}$$

are the Fourier coefficients of function f.

In terms of convergence (summability) of general orthonormal series the following well-known result of D.E. Menshov (see e.g. [1] ch. 2, #5 p. 111) is most important.

Theorem A *Let* (φ_n) *be an ONS on* $[0, 1]$. *If the series*

$$\sum_{n=1}^{\infty} |a_n|^{2-\varepsilon}$$

is convergent for some $\varepsilon \in (0, 1)$, *then the series*

$$\sum_{n=1}^{\infty} a_n \varphi_n(x)$$

is unconditionally convergence a.e. on $[0, 1]$.

On the other hand, for different classes of functions in terms of convergence(summability) of their respective Fourier series the following result of S. Banach (see e.g. [2]) is most important:

Theorem B *Let* $f \in L_2$ *be an arbitrary, non-zero function. Then there exists an ONS* (φ_n) *such that*

$$\limsup_{n\to\infty} |S_n(x, f)| = +\infty \text{ a.e. on } [0, 1],$$

where

$$S_n(f, x) := \sum_{k=1}^{n} C_k(f)\varphi_k(x)$$

and $C_k(f)$ *is the Fourier coefficients of the function* $f \in L_2$ *with respect to the system* $\{\varphi_k\}$.

We also need the following theorem (see [11],Ch.II.p.40)

Theorem C *If the series*

$$\sum_{n=1}^{\infty} a_k b_k$$

is converges for any $(b_n) \in l_q$, *then* $(a_n) \in l_p$, *where* $\frac{1}{p} + \frac{1}{q} = 1$.

From Theorem A, Theorem B, and Theorem C we can conclude that issues of convergence (summability) of general orthogonal series and general Fourier series are fundamentally different. In the first case the convergence of the series is determined by the properties of its numerical coefficients, while in the second case good differential properties of functions do not guarantee the convergence of their respective general Fourier series.

We need the following important Lemma (see [8]).

Lemma 2.1 *Suppose that* $f,\ F \in L_2$ *and* f *has only finite value in any point on* $[0, 1]$. *Then*

$$\begin{aligned}\int_0^1 f(x)F(x)\,dx &= \sum_{i=1}^{n-1}\left(f\left(\frac{i}{n}\right) - f\left(\frac{i+1}{n}\right)\right)\int_0^{i/n} F(x)\,dx \\ &+ \sum_{i=1}^{n}\int_{(i-1)/n}^{i/n}\left(f(x) - f\left(\frac{i}{n}\right)\right)F(x)\,dx \\ &+ f(1)\int_0^1 F(x)\,dx.\end{aligned} \tag{2.4}$$

3 Unconditional Convergence

The following problem arises: find the sufficient conditions to be satisfied by (ONS) functions of the orthonormal system (φ_n), so that the Fourier series of functions of a certain class be convergent. At the same time, it should be shown that the conditions that have been found are best possible and easy to verify.

Our first main result reads:

Theorem 3.1 *Let* (φ_n) *be an ONS and suppose that* $\varepsilon \in (0, 1)$ *is an arbitrary number. If* $(C_n(l)) \in l_{p(\varepsilon)}$, *where* $l(x) = 1,\ x \in [0, 1]$ *and for arbitrary* $(a_n) \in l_{q(\varepsilon)}$

$$M_n(a, \varepsilon) = O_\varepsilon(1), \tag{3.1}$$

then for any $f \in Lip_1$

$$\sum_{n=1}^{\infty} |C_n(f)|^{2-\varepsilon} < +\infty.$$

Next we state a result which, in particular, show that the statement in Theorem 3.1 is, in a sense, sharp.

Theorem 3.2 *Let* (φ_n) *be an ONS on* $[0, 1]$ *and suppose that* $\varepsilon \in (0, 1)$ *is an arbitrary number. Now let us assume that for some* $(b_n) \in l_{q(\varepsilon)}$

$$\limsup_{n\to+\infty} M_n(b, \varepsilon) = +\infty. \tag{3.2}$$

Then there exists a function $g \in Lip_1$*, such that*

$$\sum_{k=1}^{\infty} |C_n(g)|^{2-\varepsilon} = +\infty.$$

In particular, Theorems 3.1 and 3.2 imply the following new result:

Corollary 3.3 *Let* (φ_n) *be an ONS. For some* $\varepsilon \in (0, 1)$, $(C_n(l)) \in l_{p(\varepsilon)}$, *where* $l(x) = 1$, $x \in [0, 1]$ *and for arbitrary* $(a_n) \in l_{q(\varepsilon)}$

$$M_n(a, \varepsilon) = O_\varepsilon(1),$$

then for any $f \in Lip_1$

$$\sum_{n=1}^{\infty} C_n(f)\varphi_n(x)$$

is unconditional convergent a. e. on $[0, 1]$.

The validity of Corollary 3.3 derives from Theorems 3.1 and B.

Theorem 3.4 *Let* (d_n) *be a nondecreasing sequence of positive numbers. Any ONS* (φ_n) *contains a subsystem* (φ_{n_k}) *such that*

$$M_n(da, \varepsilon) = O_\varepsilon(1) \tag{3.3}$$

for any $\varepsilon \in (0, 1)$ *and* $(a_n) \in l_{q(\varepsilon)}$. *Where* $da = (d_n a_n)$.

Theorem 3.5 *Let* (d_n) *be a nondecreasing sequence of positive numbers and* $d_n = O(\sqrt{n})$. *Any ONS* (φ_n) *contains a subsystem* (φ_{n_k}) *such that the series*

$$\sum_{k=1}^{\infty} d_k C_{n_k}(f)\varphi_{n_k}(x)$$

is unconditionally convergent a.e. on $[0, 1]$ *for an arbitrary* $f \in Lip1$.

Theorem 3.6 *There exists a function $g \in Lip_1$ and ONS (Φ_n) such that*

(1) $\int_0^1 \Phi_n(x)dx = 0, \quad n = 1, 2, \ldots.$

(2) *For any $\varepsilon > 0$,*

$$\sum_{n=1}^{\infty} |C_n(g, \Phi)|^{p(\varepsilon)} = +\infty,$$

where

$$C_n(g, \Phi) = \int_0^1 g(x)\Phi_n(x)dx = 0, \quad (n = 1, 2, \ldots).$$

4 Problems of Efficiency

We call the condition $M_n(a, \varepsilon) = O(1)$ efficient if it is easily verified for classical ONS.

Theorem 4.1 *Let (φ_n) be an ONS and uniformly with respect to $x \in [0, 1]$*

$$\int_0^x \varphi_n(u)du = O(1)\frac{1}{n}.$$

Then for arbitrary $(a_n) \in l_2$

$$M_n(a, \varepsilon) = O(1). \tag{4.1}$$

Theorem 4.2 *If (X_n) is the Haar system, then for any arbitrary $(a_n) \in l_2$, the condition* (4.1) *holds.*

References

1. G. Alexits, *Convergence Problems of Orthogonal Series*, International Series of Monographsin Pure and Applied Mathematics, vol. 20 (1961)
2. S. Banach, *Theorie des operations lineaires* (American Mathematical Society, 1978), 259 pp.
3. D. Baramidze, L. Baramidze, L.-E. Persson, G. Tephnadze, Some now restricted maximal operators of Fejér means of Walsh-Fourier series. Banach J. Math. Anal. **17**(4), 75 (2023)

4. D. Baramidze, L.-E. Persson, H. Singh, G. Tephnadze, Some new weak $(H_p - L_p)$ type inequality for weighted maximal operators of partial sums of Walsh-Fourier series. Mediterr. J. Math. **20**(5), paper no. 28, 13pp. (2023)
5. D. Baramidze, L.-E. Persson, G. Tephnadze, Some new $(H_p - L_p)$ type inequalities for weighted maximal operators of partial sums of Walsh-Fourier series. Positivity **27**(3), paper no. 38, 14 pp. (2023)
6. D. Baramidze, I. Blahota, G. Tephnadze, R. Toledo, Martingale Hardy spaces and some new weighted maximal operator of Fejér means of Walsh-Fourier series. J. Geom. Anal. **34**, paper no. 3, 17 pp. (2024)
7. L. Gogoladze, G. Cagareishvili, General Fourier coefficients and convergence almost everywhere. Izv. Ross. Akad. Nauk. Ser. Math. **85**(2), 60–72 (2021); Izv. Math. 85, no.2, 228–240 (2021)
8. L. Gogoladze, V. Tsagareishvili, Unconditional convergence of Fourier series of functions of bounded variation. Siberian Math. J. **59**, 65–72 (2018)
9. L. Gogoladze, V. Tsagareishvili, Unconditional convergens of general Fourier series. Trudy Mat. Inst. imeni V.A. Steklova **319**, 83–93 (2022)
10. B.I. Golubov, A.V. Efimov, V.A. Skvortsov, *Walsh Series and Transforms* (Kluwer Academic Publishers Group, Dordrecht, 1991)
11. G. Hardy, E. Littlewood, G. Polya, *Inequalities*, 2nd edn. (Cambridge University Press, 1952), pp. xii, 324. 27s. 6d
12. S. Kaczmarz, H. Steinhaus, *Theorie der Orthogonalreihen* (Chelsea Publishing Company, 1951), 296 pp.
13. B.S. Kashin, A.A. Saakyan, *Orthogonal Series*, Translations of Mathematical Monographs (American Mathematical Society, 2005), 451 pp.
14. D. Lukkassen, L.E. Persson, G. Tephnadze, G. Tutberidze, Some inequalities related to strong convergence of Riesz logarithmic means of Vilenkin-Fourier series. J. Inequal. Appl. **2020**, paper no. 79, 17 pp. (2020)
15. N. Nadirashvili, L.-E. Persson, G. Tephnadze, F. Weisz, Vilenkin-Lebesgue points and almost everywhere convergence of Vilenkin-Fejér means and applications. Mediterr. J. Math. **19**(5), paper no. 239, 16 pp. (2022)
16. N. Nadirashvili, G. Tephnadze, G. Tutberidze, Almost everywhere and norm convergence of approximate identity and Fejér means of trigonometric and Vilenkin systems. Trans. A. Razmadze Math. Inst. **177**(3), 453–463 (2023)
17. A.M. Olevskii, Orthogonal series in terms of complete systems. Mat. Sbornik. Novaya Seriya 58 **100**(2), 707–748 (1962)
18. W. Orlicz, *Zur theorie der Orthogonalreihen* (Bull. Int. Acad. Pol. Sci., Cracovie, 1927), pp. 81–115
19. L.E. Persson, G. Tephnadze, P. Wall, On an approximation of 2-dimensional Walsh-Fourier series in the martingale Hardy spaces. Ann. Funct. Anal. **9**(1), 137–150 (2018)
20. L.E. Persson, G. Tephnadze, G. Tutberidze, On the boundedness of subsequences of Vilenkin-Fejér means on the martingale Hardy spaces. Oper. Matrices **14**(1), 283–294 (2020)
21. L.E. Persson, G. Tephnadze, G. Tutberidze, P. Wall, Some new results on the strong convergence of Fejér means with respect to Vilenkin systems. Ukr. Math. J. **73**, 635–648 (2021)
22. L.E. Persson, G. Tephnadze, F. Weisz, *Martingale Hardy Spaces and Summability of One-dimensional Vilenkin-Fourier Series*, book manuscript (Birkhäuser/Springer, 2022), 626 pp.
23. L.-E. Persson, F. Schipp, G. Tephnadze, F. Weisz, An analogy of the Carleson-Hunt theorem with respect to Vilenkin systems. J. Fourier Anal. Appl. **28**, 48, 1–29 (2022)
24. H. Rademacher, Einige Sätze über Reihen von allgemeinen Orthogonalfunktionen. Math. Ann. **87**, 112–138 (1922)
25. F. Schipp, W.R. Wade, P. Simon, J. Pál, *Walsh Series. An Introduction to Dyadic Harmonic Analysis* (Adam Hilger, Bristol, 1990)
26. K. Tandori, Uber die orthogonalen Funktionen. X. (Unbedingte Konvergenz.)? Matematika **9**, 6, 71–108 (1965); Acta Sci. Math. (Szeged), 23 (1962), 185–221

27. G. Tephnadze, On the partial sums of Walsh-Fourier series. Colloq. Math. **141**(2), 227–242 (2015)
28. G. Tephnadze, On the convergence of partial sums with respect to Vilenkin system on the martingale Hardy spaces. J. Contemp. Math. Anal. **53**(5), 294–306 (2018)
29. G. Tephnadze, On the partial sums of Vilenkin-Fourier series. J. Contemp. Math. Anal. **54**(6), 23–32 (2019)
30. G. Tephnadze, G. Tutberidze, A note on the maximal operators of the Nörlund logaritmic means of Vilenkin-Fourier series. Trans. A. Razmadze Math. Inst. **174**(1), 107–112 (2020)
31. V. Tsagareishvili, Absolute convergence of Fourier series of functions of the class $Lip1$ and of functions of bounded variation. Izvestiya Math. **76**(2), 419–429 (2012)
32. V. Tsagareishvili, On the Fourier coefficients of functions with respect to general orthonormal-systems. Izvestiya Math. **81**(1), 179 pp. (2017)
33. V. Tsagareishvili, Problems of unconditional convergence. Anal. Math. **48**, 1213–1229 (2022)
34. V. Tsagaraeishvili, G. Tutberidze, Multipliers of absolute convergence. Mat. Zametki **105**(3), 433–443 (2019); and Mat. Notes 107 (2019), no.3–4, 439–448
35. V. Tsagaraeishvili, G. Tutberidze, Absolute convergence factors of Lipshitz class functions for general Fourier series. Georgian Math. J. **29**(2), 141–161 (2022)
36. V. Tsagaraeishvili, G. Tutberidze, Some problems of convergence of general Fourier series. J. Contemp. Math. Anal. **57**(6), 367–377 (2022)
37. F. Weisz, *Martingale Hardy Spaces and their Applications in Fourier Analysis* (Springer, Berlin, 1994)
38. A. Zygmund, *Trigonometric Series*, vol. 1 (Cambridge University Press, 1959)

SPRINGER NATURE

GPSR Compliance

The European Union's (EU) General Product Safety Regulation (GPSR) is a set of rules that requires consumer products to be safe and our obligations to ensure this.

If you have any concerns about our products, you can contact us on ProductSafety@springernature.com

In case Publisher is established outside the EU, the EU authorized representative is:

Springer Nature Customer Service Center GmbH
Europaplatz 3
69115 Heidelberg, Germany

The manufacturer's authorised representative in the EU is Springer Nature Customer Service Centre GmbH, Europaplatz 3, 69115 Heidelberg, Germany. If you have any concerns regarding our products, please contact ProductSafety@springernature.com

Printed and bound by CPI Group (UK) Ltd, Croydon, CR0 4YY
01/10/2026
02225954-0015